LIFE AND THE AURA OF PERPETUAL IMPERMANENCE:

The Dark Matter Inhabiter, the Pawn, and the Normal Matter Computer Brain

Holly Marie Pollard-Wright

ISBN: 978-1-4834-9303-9 (sc)
ISBN: 978-1-4834-9302-2 (e)

Library of Congress Control Number: 2018912932

Lulu Publishing Services rev. date: 11/19/2018

To my husband, Mark, and son, Benny, for providing a solid foundation of love and support over many years that has allowed me to explore the complexities of realities without judgment. As we often tell one another, our love is infinite, and my ability to extrapolate this feeling to all living beings within the psyches is a measurement of my understanding of the true nature of reality.

Lorelei McClure, the interaction between us in this life began for me through a friendship with Michael (Mac Mac), your son, when my time inhabiting this normal matter computer brain had just begun.

You were a source of inspiration for me while growing up and have continued to be the loving presence within our psyches entangled. Yet we are both aware that we remain forever connected as part of the same universe's dark energy substrate layer, and for that I am eternally grateful.

My friend Susan Safko. Though we have never met in person, my dear brother Don provided the connection. Your support has been enduring and an extraordinary gesture of love for which I am thankful.

CONTENTS

PREFACE

Is there an all-encompassing, coherent theory that explains a deeper underlying reality for individuals with multifaceted intelligence and varied experiences? Perhaps an explanatory framework for the physical aspects of the universe is attainable, but reality experienced by an individual is unique. As such, so will his or her beliefs be unique.

From a relative perspective, the publishing of this book is simply one ordinary person taking part in the discussion of why we are here and why the universe exists. This, to me, is one of the most important questions to ask oneself before the inevitable event of brain death. Yet reality itself is shrouded in mysteries, such as why things in the phenomenal world appear the way they do. This means there are many unanswered questions surrounding what we experience as "real." I have searched for the answers using many areas of science, including neurobiology, neurology, theoretical physics, psychology, and artificial intelligence (AI). I have also used Buddhist teachings of the revered master, scholar, and teacher Dilgo Khyentse.

Over many years of thought experiments, I attempted to unify what I learned. This is not to say that my efforts will fully describe your reality, since there are aspects to it that are particular to the individual. My sincere hope is that if you are on a journey of discovery and informed decision-making, something in the pages of this book might be of benefit to you.

CLARIFICATION STATEMENT

While many individuals will formulate a relative viewpoint of reality that contains inaccuracies, this does not imply that an initial limited perspective is without merit, simply because a complete and accurate perspective of existence is obtained only by someone willing to assemble a plethora of relative viewpoints. The relative viewpoints that arise in the psyches of many can be used by each individual to formulate an accurate depiction of reality—but only if each party is willing to dispense with data that is no longer suitable and adopt new information that fits.

This book contains terms pertaining to brain structures and chemicals that have not been changed, though the relative perspective in which they were originally intended has been modified. Neuroscience explains brain function that can be extrapolated to the normal matter computer brain that embodies the inhabiter described in this book. When brain structures or chemicals are mentioned by name in this book, the reader is reminded that these are not solid but rather made of the energy called normal matter. The location of the specific brain structures mentioned is meant to imply that they are located within the normal matter computer brain that is configured to produce data that enabled imprint strings will use to create a human psyche.

The references in this book to the Wachowskis' *Matrix* movies are meant to serve as the bridge between the scientific and the theoretical. The movies take place in a world where "reality" is the product of a projection by artificial intelligence. The analogy of the *Matrix* being the psyche is a comparison that might help put the advanced concepts outlined in this book into a framework that the layperson can understand.

CHAPTER 1

A Universe of Unique Computers

There are myriad ways to occupy the universe and coexist with other energy forms, but is there a way to do so that eliminates suffering for all and results in universal bliss? In order to answer this question, a journey of discovery must be undertaken. The purpose of this exploration is to gather pertinent information, identify paths of alternative actions, and discern the consequences of these actions.[1] One can make an informed decision concerning this question only after these steps have been taken. If the answer produced is undoubtedly yes, then one must choose the path that leads to this reality. However, choice alone is not enough. It is only when it is transformed as a foothold for right action that the fruition of one's efforts is actualized.[2]

The universe is made of two types of energy: dark energy and normal energy.[3] Everything that is perceived and experienced in reality is a product of these two types of energy. When interacting with each other, both dark energy and normal energy use a different form. Therefore, both of these energies have two forms: one that is used for interacting and one that is present when not interacting. When dark energy interacts with normal energy, it does so as dark matter. When normal energy interacts with dark energy as dark matter, it is transformed to normal matter, which is visible.[4] When dark and normal energy are not interacting with each other, dark matter and normal matter are not created. When not interacting, normal energy is in an undecided state of existence and both dark energy and normal energy are invisible.[5]

Dark energy is an intelligent form of energy. Its intelligence is unmatched in the universe; it is neither created nor destroyed. It simply is.[6] Dark energy is devoid of created conscious states. It is pure awareness that produces an unprecedented, compassionate understanding of reality.[7] Combined in its two forms, dark energy constitutes 95 percent of the universe's total energy composition.[8] The universe is dynamic, however; therefore, it has the ability to change its composition. Currently, the universe is made up of 68 percent dark energy and 27 percent dark matter.[9] Collectively, both forms of dark energy are the foundation for all forms of life.[10]

The universe's vast substrate layer is made of dark energy in its two different forms. This gives the substrate layer different tendencies, based on its composition of dark matter and dark energy. Presently, all of the universe's dark matter is interspersed throughout the substrate layer as countless individual focal points that interact with the ambient energy that glides across the substrate's surface. The portion of the substrate layer that is not dark matter is the universe's dark energy—that is, a smooth, nonfluctuating stream that has no boundaries or edges.[11] This is pure dark energy that is not adulterated by interactions with normal energy, has no origination, and is unceasing,[12] giving the universe the quality of having no beginning and no moment of creation.[13]

Five percent of the universe is made of ambient energy called normal energy.[14] Normal energy occupies the universe most commonly in its visible form (normal matter) while interacting with the universe's existing focal points (dark matter). What drives normal energy to form normal matter is dark energy's desire to engage with it in a meaningful way. The current state of the universe's energy composition is due to the intelligent design of dark energy. When dark energy initiates the interaction, it outnumbers normal energy by a large percent. Therefore, rather than engaging normal energy all at once, dark energy does so slowly, in a purposeful way, in its form of dark matter. Until all of the dark matter contained within the universe's substrate layer has interacted with normal matter, the interaction between these two energies will continue. That is not to say that normal energy will always reside in the dynamic universe as normal matter; it will not. As individual focal points of dark matter complete the journey, they will be permanently changed to the dark energy that does not interact with normal matter. However, the transformation of one individual focal point of dark matter is an imperceptible amount of change, and the momentary occurrence where normal energy is not in its visible form is brief and undetectable. As the process of transformation of these two energies continues, the universe's energy composition will change to the point that there is more dark and normal energy and less dark and normal matter.

Intelligence is an important distinguishing feature when comparing dark energy to normal energy. Dark energy's intelligence is not created but has a natural self-consciousness that lies beyond subject-object duality.[15] The consciousness of dark energy is one that never changes and is devoid of the concepts that are present within a created psyche.[16] Dark energy's intelligence does not include the limitations of created conscious states, and it is not ensnared by illusionary appearances.[17] Normal energy has no inherent intelligence, but in its visible form and while interacting with dark matter, normal matter transforms the energy fields present in the universe to waves of electrical patterns that can be described as a kaleidoscope of waves of different kinds.[18] All observable forms in the universe are a result of the conscious states created when normal matter is unified into a unit that functions collaboratively with dark matter through a complex network of integration. However, a fabricated consciousness is fragile in nature and must be created constantly; therefore, this consciousness is in an ever-changing flux.[19] It is the differences in intelligence that create the reality that dark energy and normal energy have opposite effects on space.[20]

There is only one stable form of intelligence in the universe capable of responding to a created psyche in a way that results in a metaphysical transformation, and that is dark energy. Unlike normal energy, dark energy's intelligence can freely choose its actions because it is unconstrained by the conscious states that need to be created to exist and are unstable and subject to change.[21] What dark energy is not free to evade is the universal law of causality, where for every decision it makes and resultant action taken, there is a consequence.[22]

Out of the universe's total energy composition, 68 percent has undertaken an odyssey of discovery and is consequently transformed to a smooth, noninteractive stream of dark energy.[23] This is the action determined by the compassionate intelligence of the universe's dark energy, whose decision was made after immersing itself in the world of normal matter. It immersed itself because although applicable information could

3

be obtained through self-assessment, dark energy understood that this would only yield a portion of the needed relevant information.[24] Although in comparison normal energy is a much smaller percentage of the universe's overall energy composition, size does not dictate the limit of dark energy's compassion. In order to overcome any barriers that might hinder its efforts to live harmoniously with normal energy,[25] dark energy ascertained that it must engage normal energy to attain vital information.[26] Dark energy has been slowly engaging the normal energy with which it shares the universe.

Key Point

- Everything that constitutes your reality is, in actuality, energy forms interacting.

The universe's dark energy is a nonfluctuating stream of energy with no boundaries or edges, and the ambient energy is normal energy that glides across the dark energy substrate surface.

Normal energy glides across the dark energy substrate surface.

Reminders before Reading Chapter 2

- dark energy: One of the two types of energies present in the universe, which is an intelligent form of energy that is not created but has a natural self-consciousness that lies beyond subject-object duality.
- dark matter: The form dark energy uses when interacting with normal matter.
- normal energy: One of the two types of energies present in the universe, lacking natural consciousness.
- normal matter: The form normal energy uses when interacting with dark matter.
- universe: Made of two types of energy (dark energy and normal energy) and contains the interactive form of these two energies (dark matter and normal matter, respectively).
- universe's substrate layer: Made of pure dark energy that is a nonfluctuating stream of energy with no boundaries or edges.

CHAPTER 2

The Odyssey of Engagement

Because there was never a time where there was not dark energy and normal energy, the universe was not "created" as if the universe actually came into existence, exists, and will cease to exist.[27] What began was the synergy between dark energy and normal energy. Prior to the interaction, these energy types were not interacting. Therefore, the universe was invisible. The universe began to appear when dark energy made a decision to change from its form as a nonfluctuating stream to a focal point of dark energy that fluctuates. The fluctuations that occur on the surface of a focal point of dark energy are its distinguishing characteristic. It distinguishes dark energy in its form as dark matter from that of its form as a nonfluctuating stream. When dark energy assumes its form of dark matter and interacts with normal matter, the visible forms in the universe appear.

After a focal point of dark matter gathers enough pertinent information to identify the paths of alternative actions and discern the consequences of those actions,[28] it will make an informed decision and can choose to revert back to its original form. In this way, the current composition of dark energy is a recording that reverberates a long history of an ongoing dialogue between the two energies that share the universe. The universe's nonfluctuating dark energy stream represents conversations concluded, whereas individual focal points of dark matter are part of the current ongoing conversation, with each focal point having a unique timeline of when the conversation will end. What causes the fluctuations on the surface of a focal point of dark energy are generated from filaments of dancing energy.[29] These dancing filaments of energy are called imprint strings that interact with normal matter, and arising[30] from this event is the birth of the inhabiter. An inhabiter is one of many countless dark-matter focal points within the universe's substrate layer that came into existence fluctuating with a surface of vibrating imprint strings.[31]

When dark matter and normal matter interact, a created consciousness is conceived, yet this consciousness will contain inaccuracies. An inhabiter will hold a true picture of reality when it relies on its dark-energy intelligence to guide it in the creation of the psyche it fabricates with normal matter. Therefore, the inhabiter must have a connection to its previous form as a nonfluctuating stream of dark energy, and this reveals the intelligence of the inhabiter's design. The inhabiter is configured to include a focal point of dark energy that is its intelligence, which exceeds that constructed with normal matter. This acts as a base of support to the inhabiter that skillfully uses it; but the law of causality dictates that for every decision an inhabiter makes, there is a resultant consequence.[32] If the inhabiter uses its base of support adeptly, an accurate view of reality will result; if it does not, the inhabiter will experience the consequence of that action and will not perceive reality as it truly exists.

One inhabiter is a very small portion of the universe's 27 percent[33] of dark matter that is embarking on its journey to informed decision-making. When a focal point of dark energy develops a surface of vibrating imprint strings, the universe's substrate layer and normal energy mark this event with a corresponding response. The oscillating imprint strings demarcate the peripheral dimensions of a focal point of dark energy and distinguish the focal point from the universe's dark energy substrate layer. When part of an inhabiter's configuration, dark energy as a focal point is still attached to the universe's dark energy substrate layer but has on its peripheral surface oscillating imprint strings. Imprint strings are the instrument dark energy has chosen to interact with normal matter in a meaningful way, and are discrete quantities of energy that represent a metamorphosis from the universe's dark energy substrate layer to a focal point of dark matter. The inhabiter—now differentiated from the universe's dark energy substrate layer via its surface of oscillating imprint strings—communicates a gravitational force to the normal energy close by the focal point of dark matter.[34] This gravitational force is so strong[35] that any normal energy near its boundary is pulled toward the focal point of dark matter, and when pulled near, normal energy folds and forms as normal matter around the inhabiter.

How normal matter folds around the inhabiter creates the shape and structure of the normal matter computer brain. When forming a computer brain with the capacities to compute for a human psyche, normal matter curls upon itself like a tent pulled by ropes with a pole in the middle,[36] and normal matter is formed to connect all parts of what is to become the normal matter computer brain. Collectively, the transformation of normal energy to normal matter forms the human computer brain, a unique and powerful computer made of normal matter that is situated on the universe's dark energy substrate layer.

The inhabiter and the normal matter computer brain form a singular point[37] within the fabric of the universe, in which all of the energies contained within them are unified into one configuration. This focal point of unification produces different levels of consciousness: alertness with awareness, alertness and awareness with a history, and one that has memories of the past and plans for the future.[38] Spawned by the interaction of the inhabiter with the normal matter computer brain, these conscious states generate the concepts of space and time, memory, reasoning, imagination, creativity, and language.[39] The information created within these conscious states will vary between inhabiters and is dependent on the law of cause and effect. What the inhabiter does affects the normal matter computer brain and vice versa. The output of this interaction culminates to produce the information used to generate each conscious state. Within the long history of the universe, the quantity of inhabiters that have been born in this way is innumerable and may exceed that of the number of visible stars.[40]

A focal point of dark matter and the normal matter that surrounds it will remain connected for an unrevealed period of time that might allow the inhabiter to complete its odyssey of discovery. As one inhabiter completes its journey, a signal is echoed throughout the universe's substrate layer, indicating that another focal point is to begin transforming itself to dark matter by developing a surface of oscillating imprint strings. The inhabiter that completed its journey will no longer be communicating with normal matter and will therefore not need its imprint strings; they are then absorbed by the focal point of dark energy and disappear.

When this happens, what was once a focal point of dark matter as an inhabiter becomes a nonfluctuating stream of dark energy, as there is no differentiation between the substrate layer and the focal point; it's as if a demarcated focal point is smoothed away.

Key Point

- The relationship between the inhabiter and the computer brain is special because of the way these two energy forms interact and the fact that the inhabiter is the only observer of space and time.

An inhabiter is a focal point of dark matter, which is a focal point of the universe's dark energy with a surface of fluctuating imprint strings.

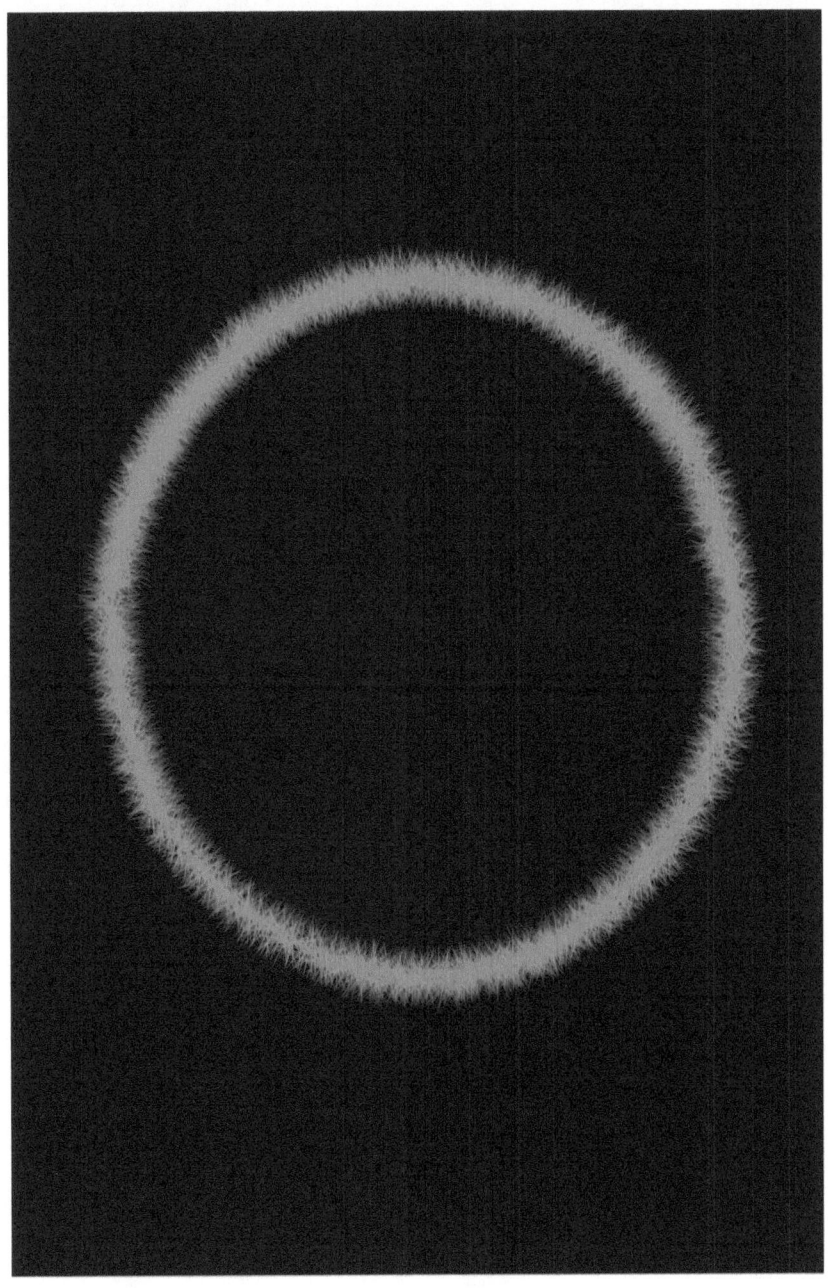

A dark matter inhabiter.

The dark matter inhabiter is differentiated from the universe's dark energy substrate layer via its surface of oscillating imprint strings and communicates a gravitational force to the normal energy close by the focal point of dark energy. This gravitational force is so strong that any normal energy near its boundary is pulled toward the focal point of dark energy, and when pulled near, normal energy folds and forms as normal matter around the inhabiter.

How normal matter folds around the inhabiter creates the shape and structure of the normal matter computer brain. When forming a computer brain with the capacities to compute for a human psyche, normal matter curls upon itself like a tent pulled by ropes with a pole in the middle, and normal matter is formed to connect all parts of what is to become the normal matter computer brain.

Normal matter folds around the inhabiter to create a computer brain.

Reminders before Reading Chapter 3

- created consciousness (psyche): A product of dark matter and normal matter's interaction.
- focal point of dark energy: It is within the inhabiter's configuration, derived from the universe's substrate layer; that is, the inhabiter's intelligence that exceeds the created conscious states that the inhabiter creates while interacting with normal matter.
- focal point of dark matter: A focal point of the universe's dark energy substrate layer with a surface of imprint strings.
- imprint strings: Dancing filaments of energy on the peripheral surface of the inhabiter's focal point of dark energy that differentiate the inhabiter from the universe's dark energy substrate layer. Imprint strings oscillate and produce a gravitational force.
- inhabiter: A focal point of dark matter, which is a focal point of the universe's dark energy with a surface of fluctuating imprint strings.
- normal matter computer brain: Consisting of the normal energy that was drawn to the inhabiter by the gravitational force produced by the inhabiter's oscillating imprint strings, normal energy combined and thereby was transformed to the normal matter computer brain that embodies the inhabiter.
- psyche: Created conscious states that are produced when dark matter and normal matter interact.
- universe: Was not created but was invisible until dark energy and normal energy began interacting.
- universe's dark energy substrate layer: Represents interactions between dark matter and normal matter that have concluded.

CHAPTER 3

A World Created by Electrical Patterns

When a focal point of the universe's dark energy substrate layer is encircled by a surface of imprint strings that oscillate, a gravitational force is produced, and this draws normal energy to a region within the universe. Immediately upon being pulled toward the focal point of dark energy by oscillating imprint strings, normal energy is transformed to normal matter, and it is now the computer brain that embodies the inhabiter. Once formed, the normal matter computer brain will connect to other computer brains nearby and will share information, forming an interconnected network of normal matter computer brains. Exactly how data is relayed throughout the network of connected normal matter computer brains is perplexing, as there is no assuredness to the direction of information flow or how it will be received. Nonetheless, the inhabiter uses the normal matter computer brain's information, by means of its imprint strings, to create the psyche that is replete with uncertainty. An interesting facet of the psyche is this: even though the inhabiter's imprint strings utilize the computer brain's data to create the psyche, the psyche is not recognized by the normal matter computer brain. The reason for this is that the computer brain is normal matter, which is a form of energy that does not perceive with conscious states.

There are many different configurations of normal matter computer brains that embody an unknown number of inhabiters. However, when normal matter computer brains are not configured similarly, they will function differently. Which is to say that not all normal matter computer brains produce the same data, and therefore not all inhabiters embodied within will create the same psyche. There are computer brains of many dispositions spread throughout the universe, and each inhabiter embodied within will use the data produced by normal matter to create awareness that varies in scope. Notwithstanding the psyche that the inhabiter creates is dependent on the information the computer brain receives and normal matter's capacity to generate and process information. Computer brains of very different configurations indeed share data, but functioning variously, the information exchanged may not be recognized, or in some cases, it will be rejected, and this will affect information flow as it travels throughout the network of connected normal matter computer brains. Be that as it may, unlike normal matter computer brains that are limited by proximity and must be connected to the network of computer brains in a way that allows them to share data, focal points of dark matter are connected as part of the universe's dark energy substrate layer. All focal points of dark energy within inhabiters' configurations are forever connected to the dark energy substrate layer; that is the background energy woven into the fabric of the universe.

While embodied and as a consequence of perceiving the psyche, the inhabiter will formulate its relative view of reality, and that viewpoint will contain inaccuracies. With a relative viewpoint, some inhabiters will understand the universe as being split into two distinct regions known as the microscopic and macroscopic universe. The microscopic and macroscopic universe are the small and large dimensions of the universe respectively and are conceptualized by an inhabiter with an unreliable comprehension of the universe as having a demarcation that splits the universe into two distinct regions that do not actually exist. In truth, the universe is comprised of interacting and interconnected energies that do not function separately but as one, which means the microscopic and macroscopic universe is an illusory delineation within the psyche. Thus, this nonseparable universe can be envisioned as functioning as one. Normal matter that embodies each inhabiter forms its structure. Inhabiters with their focal point of dark energy and imprint strings are the mind embodied, and the dark energy substrate layer provides the foundation, allowing both universes to function as the one. Albeit, with the psyche, the inhabiter can conceptualize a universe that is not actually seen, and when the inhabiter mentally visualizes the universe, this suggests that the inhabiter has a relative viewpoint of reality. However, the ability to engender a picture of the universe with structure as if it is solid when it is not is useful because it helps the inhabiter formulate an understanding of its reality.

While embodied within a normal matter computer brain, what the inhabiter perceives as its reality is present only within the psyche, and the inhabiter will describe its reality symbolically and with integrated pictures, but when the inhabiter is not fully apprehending what it is or where it is, the inhabiter's relative viewpoint will be deceptive. An example of a deceptive relative viewpoint would be as follows: The inhabiter's assumption is that there are black holes present within the universe and that escape is nearly impossible for anything once trapped inside. What goes on in this imagined black hole's interior is unknown to the outside world. Yet black holes are allegorical and are present only within the psyche created by the inhabiter using its enabled imprint strings so as to conceptualize its very existence while inhabiting a normal matter computer brain. While embodied and when the inhabiter is struggling with the psyche in a way that its reality is unanchored, in this state the inhabiter's escape from conditioned existence is all but impossible. Yet what the inhabiter experiences it must share for it to be known, as the psyche is abstruse, not graspable, and cannot be directly perceived by other inhabiters.

A second common misleading relative viewpoint consists of the inhabiter's belief that it has a brain that occupies a space in its skull, when in actuality it is the inhabiter that is embodied by the normal matter computer brain. Be that as it may, a relative viewpoint that includes the brain as an organ of soft nervous tissue contained in a skull might be useful, and the brain conceptualized to function like a computer might also be of benefit. However, the normal matter human computer brain out-functions any manufactured computer.

The normal matter computer brain transforms the energy fields present in the universe to waves of energy in the form of electrical patterns of different kinds. The computer brain is an integrated electrical system, and normal matter will transmit information as waves of electrical patterns. The uppermost region and lower regions of the computer brain are separated but connected and are respectively known as the brain and body, although collectively referred to as the normal matter computer brain. Relatively, when thinking of the brain that functions like a computer, the familiar parts—monitor, mouse, and internal circuitry—are replaced by organic tissue. The brain is composed of highly specialized tissue made up of some of the billions of cells in the body. Although the brain is made of many types of cells, there is one category of cells that is imperative to generate the data that imprint string will use to create the inhabiter's conscious states. This particular cell type, an electrical cell, is called a *brain neuron,* and it is estimated there are eighty-six billion[41] of them in the brain of a normal matter computer brain that is configured to produce data that imprint strings will use to make a human psyche. Collectively, these cells form the circuitry of the brain.[42] Each brain neuron generates electrical pulses that travel the entire length of the neuron until they reach the end (or terminal). Once the electricity reaches the nerve terminal, chemicals called *neurotransmitters* are released into the synaptic cleft. The neurotransmitters are what enables one neuron to transmit its electrical impulse to another. Neuron terminals lie very close to dendrites—or branches—of other neurons. A nerve terminal, the dendrite of an adjacent neuron, and the space, or cleft, between these two things constitute one of the multitude of synapses[43] that form the circuitry of the brain.

Once integrated as a computer brain, normal matter will transmit information through electrical patterns. However, the inhabiter will interpret the information generated by the computer brain not as electrical patterns but as moving pictures that are within the psyche, using the medium of its *imprint strings.* Imprint strings oscillate and interact with the brain's electrical patterns; by doing so, they reveal to the inhabiter the many appearances of reality. Depending on the magnitude of the inhabiter's use of its imprint strings to interface with the brain's neurons, which fire at different frequencies, reality while embodied will be displayed differently. There will be versions of reality within the psyche that the inhabiter will find reassuring because they reveal an identifiable presence within the universe and a certain future. These familiar appearances bring security; for this reason, many inhabiters will cling to them, even though they are not an accurate representation of reality. A genuine view of reality is unfamiliar, and it is not certain if this viewpoint will be envisioned, because it depends on the inhabiter using its imprint strings skillfully within the laws of neural functionality. To acquire any skill requires diligence, and the inhabiter must train itself to become accustomed to unfamiliar perceptions of reality where nothing is really as it seems and everything is possible.

Key Point

- While inhabiting a computer brain, your reality is within the psyche created by your enabled imprint strings transforming electrical patterns into moving pictures. It is similar to how movement in a movie is created by a rapid succession of individual still images.[44]

The regions of a normal matter computer brain: The brain is located within the proximal region of the normal matter computer brain and functions as the coordinating center, allowing the computer brain to generate information in the form of electrical patterns. The body is the lower region within the normal matter computer brain that embodies the inhabiter.

In this picture, a "peephole" has been created, allowing the reader to peer into the interior of a normal matter computer brain that embodies the inhabiter and is configured to produce data for a human psyche.

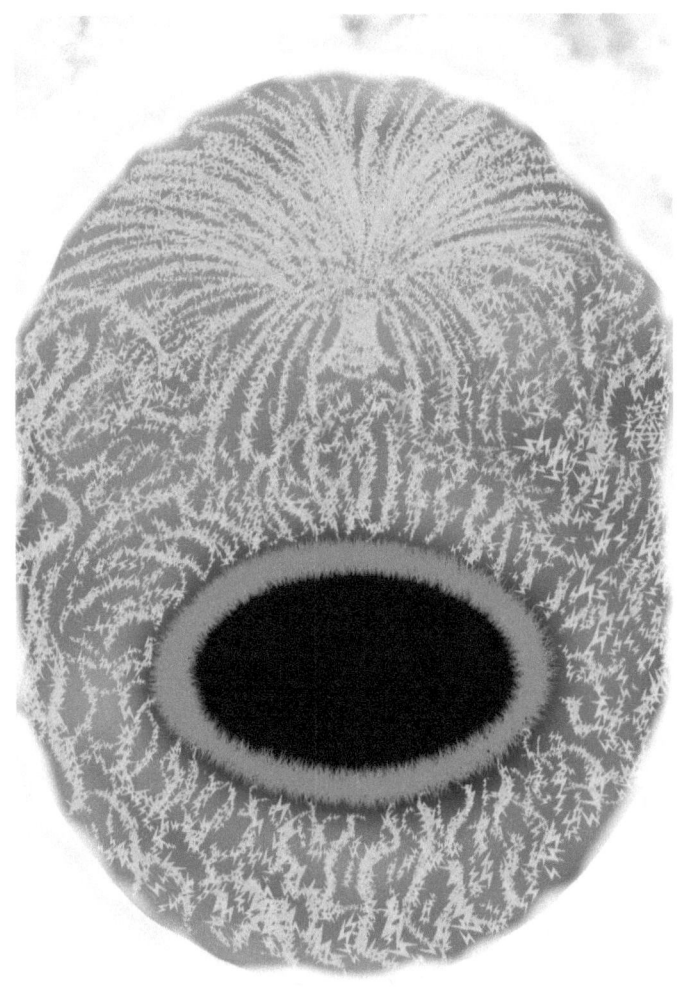

"Peephole" revealing the interior of a normal matter computer brain.

A neuron is an electrical cell and the means by which normal matter is able to generate data and communicate this information throughout the computer brain.

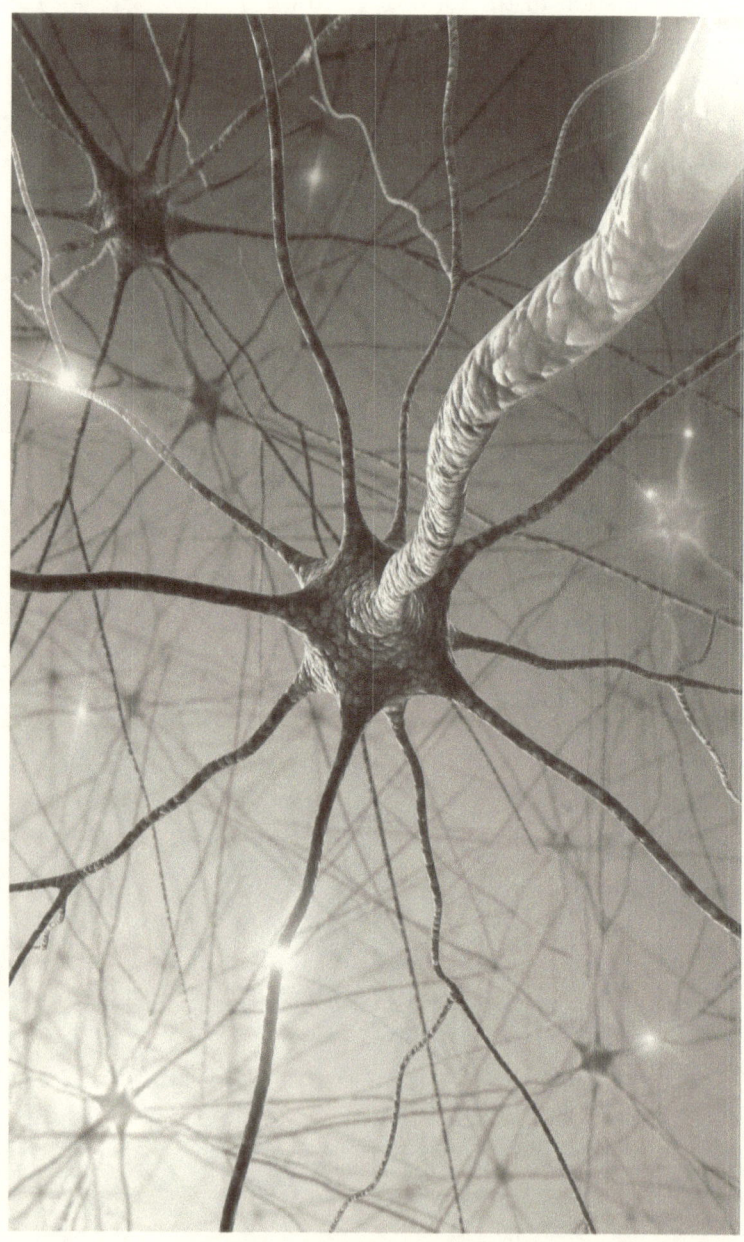

Neuron.

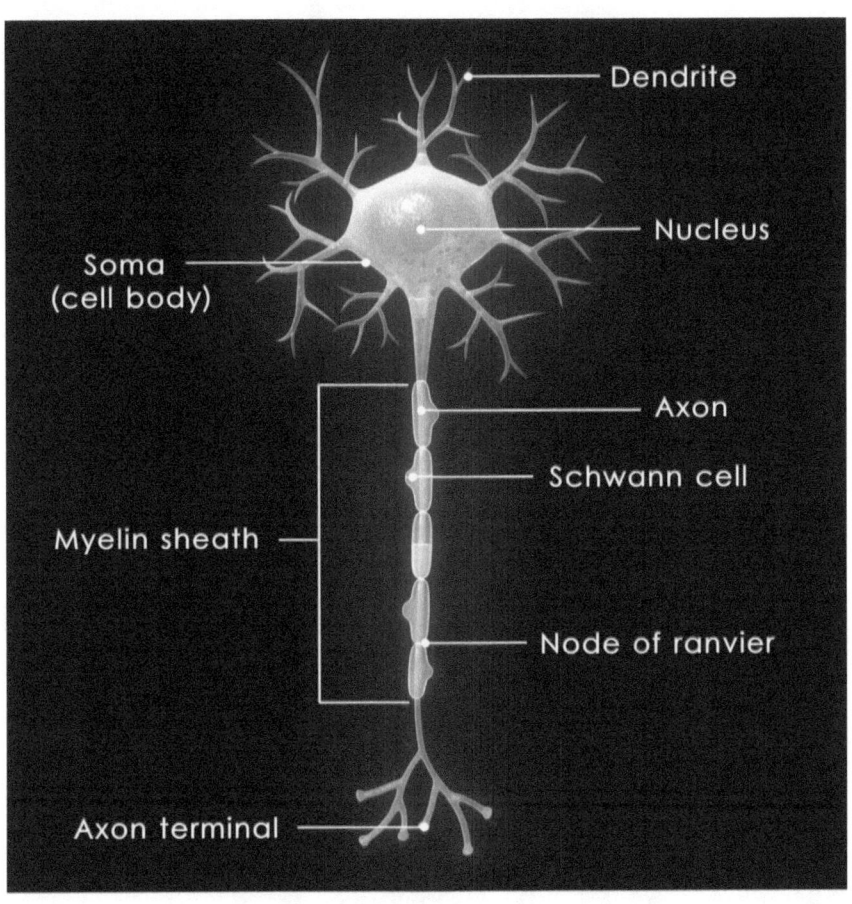

Parts of a neuron.

Neurotransmitters are chemical programs that allow one neuron to transmit its electrical impulse to another, and there is more than one kind generated by the normal matter computer brain.

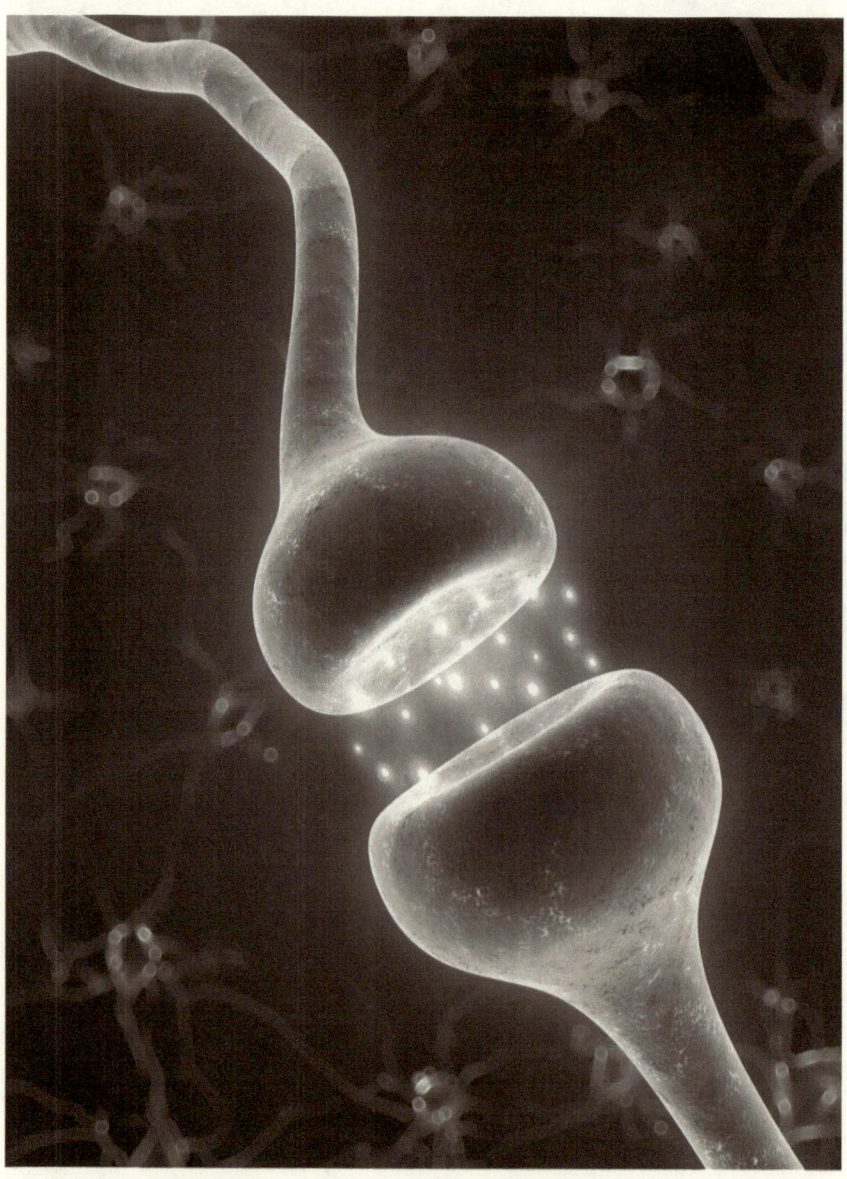

Neurotransmitters.

Normal matter computer brains that are spread throughout the nonseparable universe are connected to other computer brains nearby and relay information to one another. However, there is no assuredness to the direction of information flow throughout the network of interconnected normal matter, and there are many different kinds of configured computer brains (i.e., configured to produce data for a human being psyche versus some other living being).

In this picture, a "peephole" has been created, allowing the reader to peer into the interior of a normal matter computer brain configured to produce data for a human psyche.

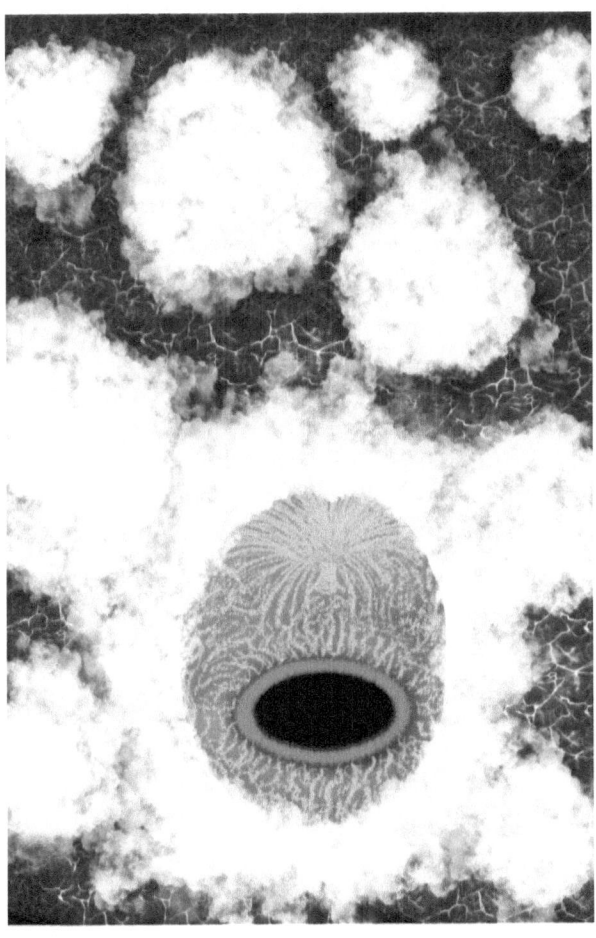

"Peephole" revealing the interior of only one normal matter computer brain that is connected to other computer brains nearby.

Reminders before Reading Chapter 4

- black holes: A figurative representation within the psyche of "reality" that the inhabiter perceives with its created conscious states and represents the inhabiter lost within a psyche while embodied within a normal matter computer brain.
- body: The lower region located within the normal matter computer brain that embodies the inhabiter.
- brain: Understood by the inhabiter, with an inaccurate relative viewpoint, to be an organ of soft nervous tissue situated within its skull. However, when the inhabiter understands the brain as being located within the proximal region of the normal matter computer brain that embodies the inhabiter and functions as the coordinating center, allowing the computer brain to generate data imprint strings will use for sensations, intellectual, and nervous activity within the psyche, then the inhabiter has an updated and more accurate relative viewpoint.
- dendrite: The branches of a neuron.
- electrical patterns: Generated by the normal matter computer brain's neurons that contain information.
- imprint strings: Utilize the computer brain's electrical patterns to create the psyche of created conscious states the inhabiter will use to perceive its "reality" displayed as moving pictures.
- interconnected network of computer brains: Made of connected normal matter computer brains that relay information among one another, albeit there is no assuredness to the direction of information flow throughout the network of interconnected normal matter computer brains.
- macroscopic universe: The large dimensions of the universe within the psyche, recognized by the inhabiter with a relative viewpoint.
- microscopic universe: A conceptual image within the psyche of the small dimensions of the universe, perceived by the inhabiter with a relative viewpoint.

- neuron: An electrical cell that generates electric pulses that are the means by which normal matter is able to generate data and communicate this information throughout the normal matter computer brain.

- nerve terminal: The end of a neuron.

- neurotransmitters: Chemical programs that allow one neuron to transmit its electrical impulse to another, and there are more than one kind generated by the normal matter computer brain.

- normal matter computer brain: Made of the energy normal matter and connects to other computer brains in the universe that are nearby, which may or may not produce the same data, depending on how the computer brain is configured (i.e., configured to produce data for a human psyche versus some other living being). The computer brain does not perceive with created conscious states.

- psyche: Made up of the created conscious states the inhabiter's imprint strings formulate by utilizing the computer brain's data and that are experienced by the inhabiter and not the normal matter computer brain.

- relative viewpoint: A viewpoint adopted by the inhabiter in which it understands "reality" and contains symbols and integrated pictures within the psyche imprint strings created with the computer brain's data.

CHAPTER 4

Imprint Strings

As they collectively make up the 27 percent[45] of the universe that is dark matter, inhabiters must embark on their own journey of decision-making.[46] The mission of each inhabiter is to ascertain how to coexist with normal matter—an energy form that is very different from itself—in a way that is mutually beneficial. Dark energy and normal energy communicate differently. The language that dark energy uses to communicate with itself would not be useful to use with normal energy. Therefore, in order to communicate, dark energy created its imprint strings so the portion of itself on exploration could make the most of its experience.

The dancing strings[47] of energy that cause fluctuations on the surface of a focal point of dark energy are the inhabiter's imprint strings. These strings are what interface with normal matter, allowing the inhabiter to interpret the electrical patterns of the computer brain. When the inhabiter uses imprint strings, its semiconscious and conscious states within the psyche are created.

The inhabiter's imprint strings have the capacity to transform electrical patterns generated by the circuitry of the brain into physical sensations known as sense impressions and mental events that are the thoughts, emotions, memories present within the psyche, allowing the inhabiter to comprehend the normal matter computer brain's electrical patterns. The inhabiter has two types of imprint strings: ego-clinging and non-ego-clinging. When starting its journey, the inhabiter will have an equal number of both types of imprint strings.

Dark energy created ego-clinging imprint strings for the purpose of allowing the inhabiter an immersive experience into the world of normal matter[48] as it is being sequenced by the normal matter computer brain. These strings interface with the brain's electrical patterns and create interpretations of these patterns by converting them into sense impressions and mental events according to how the brain guides them, and in this way, an unadulterated consciousness with normal matter is created. When using these imprint strings, the inhabiter will have an "insider" experience through a psyche called "psyche self-centered," and the inhabiter will perceive the world as normal matter creates it.

The normal computer brain arranges electrical patterns in a manner conducive to neuronal functioning. This is the computer brain's driving force for the sequencing of electrical patterns, and it is adopted by ego-clinging imprint strings to guide the sequencing of sense impressions and mental events. The inhabiter, using its ego-clinging imprint strings, will create its semiconscious and conscious states within psyche self-centered, according to the computer brain's agenda. The inhabiter, using the computer brain's data, makes these conscious states with an agenda of benefitting normal matter's neuronal circuitry. The inhabiter is not recognized as something separate from the normal matter computer brain; its future interests are not included in the brain's sequencing of electrical patterns. Conscious states created by ego-clinging imprint strings are predatory in nature. This leads the inhabiter, using its ego-clinging imprint strings, to create conscious states that obscure the innate radiance and purity of its focal point of dark energy that lies beyond subject-object duality. This results in the inhabiter responding to the psyche in a way that increases disorder in the universe.[49]

Non-ego-clinging imprint strings also interface with the computer brain and create interpretations of these patterns by converting them into sense impressions and mental events. However, dark energy created these imprint strings to fulfill a multifaceted purpose that differs from the reason it created ego-clinging imprint strings. Consequently, these strings interpret the computer brain's data differently from ego-clinging imprint strings, to create "psyche altruistic," which contains sense impressions and mental events. One purpose of non-ego-clinging imprint strings is that of enrichment.[50] By perceiving and interpreting the brain's electrical patterns, the inhabiter, using its non-ego-clinging imprint strings, gains insight into the mental and emotional processes dictated by normal matter.

Clarity of the inhabiter's perception is another reason dark energy created non-ego-clinging imprint strings. They interact with normal matter, but unlike ego-clinging imprint strings, when the inhabiter uses these strings, the inhabiter is not so immersed in the interactive experience with normal matter that a true picture of reality is absent.[51] The ability that non-ego-clinging imprint strings have to empower the inhabiter is derived from the guidance the imprint strings receive in how to interpret the computer brain's data to create sense impressions and mental events. Non-ego-clinging imprint strings do not rely on the computer brain to guide them but instead use the inhabiter's base of support, namely its focal point of dark energy, the focal point of dark energy present in every inhabiter's configuration[52] that is created from the substrate layer of the universe. It is this pure consciousness, independent of the brain, into which non-ego-clinging imprints tap. By tapping in, non-ego-clinging imprints are shaping their interpretation of the normal matter computer brain's activity without becoming ensnared by appearances. When comparing the appearance of a non-ego-clinging imprint sting to that of an ego-clinging imprint string, there are times when they both look the same, oscillating strings of energy. However, when non-ego-clinging imprint strings tap in, they bend as the means by which they receive guidance from the inhabiter's focal point of dark energy. When bent, non-ego-clinging imprint strings utilize dark energy's consciousness, which exceeds the capacities of the computer brain and is not constrained like the brain by the rules to which the computer brain must adhere in order to function, which are, in part, created by its neurons. Where there are neurons, there are rules that dictate the neuron's life span, connections, and how the neurons communicate with one another.[53] The nonfluctuating dark energy substrate layer of the universe is not bound by these limiting parameters, and this is what makes it so powerful. By utilizing this power, non-ego-clinging imprint strings have the ability to transform the electrical patterns of the computer brain in a way that allows the inhabiter to avert becoming deceived by the appearances within the psyche.[54] An inhabiter using these strings will perceive its conscious states as being generated by two forces: a normal matter computer brain and providence gleaned from the universe's dark energy substrate layer.[55] Transforming electrical patterns in this way, an Inhabiter using its non-ego-clinging imprint strings will bend and break the rules governing its creation of semiconscious and conscious states. When this happens, the inhabiter will experience deeper dimensions of its reality and see meaning hidden from an inhabiter that is only using its ego-clinging imprint strings. An inhabiter using its non-ego-clinging imprint strings will perceive itself as being separate from the computer brain but also necessarily partnering with normal matter to create reality. Using these strings, an inhabiter will take action that is diametrically opposed to the inhabiter's response to the psyche when using its ego-clinging imprint strings.

There is uncertainty[56] surrounding which of its imprint strings the inhabiter will primarily use to experience the reality of its conscious states. Didactic in its approach to interfacing with the inhabiter, normal matter guides by dictating. The inhabiter's ego-clinging imprint strings will be guided by the computer brain to sequence sense impressions and mental events in a way that allows limited means for the inhabiter to ultimately benefit itself.[57]

The guidance given by dark energy is different from that given by normal matter. Rather than dictating exactly what the inhabiter's response to the psyche should be, dark energy encourages an inhabiter toward sustained conscious states of altruism.[58] With altruism as a foundation, the inhabiter, using non-ego-clinging imprint strings, is free to explore the world of normal matter while fully experiencing the ramifications of its decisions.[59]

What is certain is that as the inhabiter uses its imprint strings, it will disable an imprint string of the opposite type. Specifically, an ego-clinging imprint string disables a non-ego-clinging imprint string, and the opposite process is at work, where a non-ego-clinging imprint string disables an ego-clinging imprint string. When an imprint string is enabled, it means that its activity is "turned on" and will actively influence the inhabiter's perception. Conversely, when disabled, an imprint string is not broken or destroyed; it is simply "turned off," and its ability to function disappears unless reactivated. Therefore, the inhabiter's imprint string population is constantly in flux, and when enabled, they influence the inhabiter's perception.[60]

Once activated, an imprint string's influence will not simply disappear; it will remain within the inhabiter's perception. Like a record, the inhabiter's imprint string population tells the story of how an inhabiter spent its time inhabiting a computer brain.[61] Each enabled imprint string acts as a promissory note[62] with regard to the inhabiter's future happiness or suffering. The exact number of imprint strings present on the surface of the inhabiter's focal point of dark energy is incalculable and unknown. This is one of the many mysteries of life embodied within a normal matter computer brain that is situated on the universe's dark energy substrate layer. Although not an indicator of number, as imprint strings are enabled and disabled while interfacing with the brain's electrical patterns, electromagnetic radiation is produced and emitted within the normal matter computer brain.[63]

Key Point

- Your ego-clinging imprint string population that is enabled constrains[64] your ability to make the choices that benefit "You," normal matter, and the nonseparable universe.

Close-up view: Non-ego-clinging imprint strings bend so they tap in and connect to the inhabiter's focal point of dark energy that guides them into creating sense impressions (physical sensations) and mental events (thoughts, emotions, memory) of the psyche altruistic, by transforming the electrical patterns generated by the circuitry of the normal matter computer brain. When non-ego-clinging imprint strings are bent and tapped in to the inhabiter's focal point of dark energy, they vibrate in different patterns.

The image displays the different vibrational patterns of non-ego-clinging imprint strings depicted stationary and vibrational.

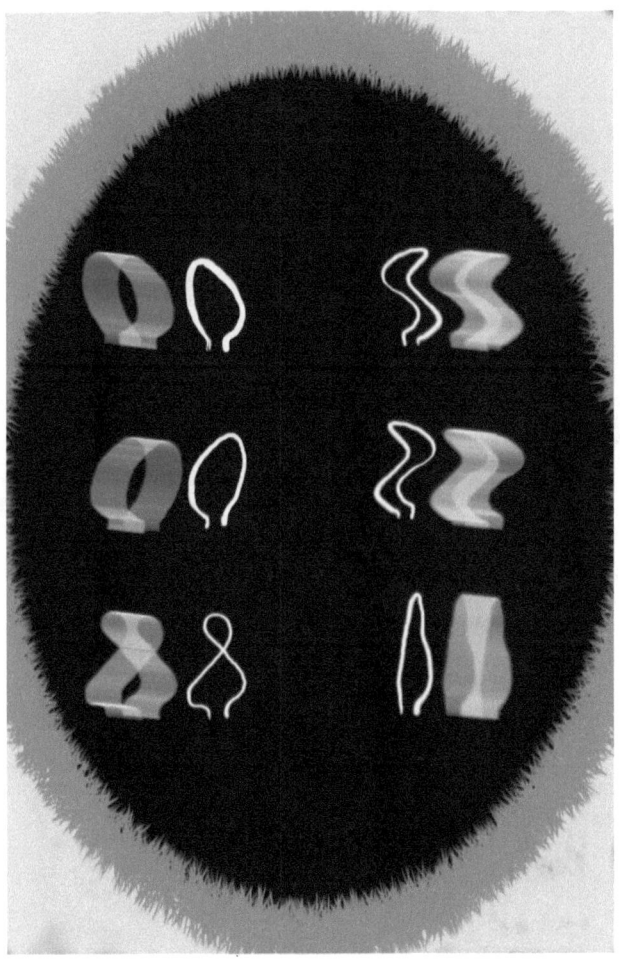

Stationary and vibrational patterns of non-ego-clinging imprint strings.

Close-up view: Ego-clinging imprint strings create the psyche self-centered, which consists of sense impressions (physical sensations) and mental events (thoughts, emotions, memory) by transforming the electrical patterns generated by the circuitry of normal matter according to how the computer brain dictates. Although ego-clinging imprint strings do not tap in to the inhabiter's focal point of dark energy, they are attached to the focal point at one end and vibrate in different patterns.

The image displays the different vibrational patterns of ego-clinging imprint strings depicted stationary and vibrational.

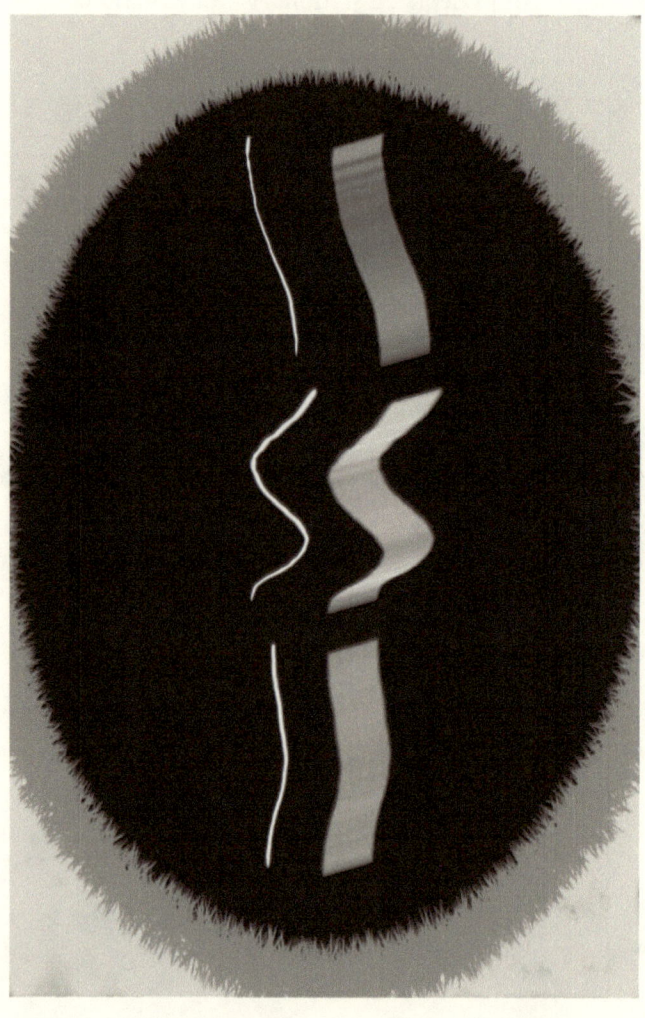

Stationary and vibrational patterns of ego-clinging imprint strings.

Oscillating, enabled imprint strings are arranged as pairs of ego-clinging and non-ego-clinging imprint strings on the inhabiter's focal point of dark energy.

There are many different vibrational pattern combinations of ego-clinging and non-ego-clinging imprint string pairs, and the image displays some of these combinations in vibration.

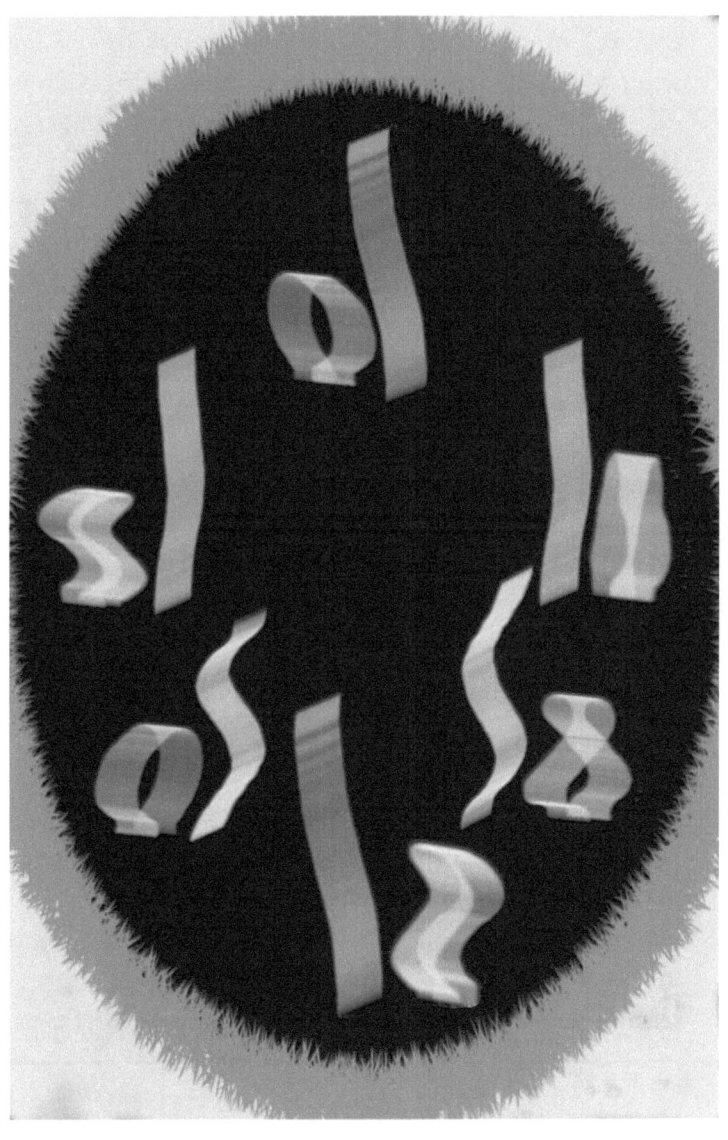

Vibrating ego-clinging and non-ego-clinging imprint string pairs.

Reminders before Reading Chapter 5

- ego-clinging imprint stings: Create psyche self-centered that is predatory in nature and are sense impressions and mental events of conscious states created by these strings, transforming the computer brain's electrical patterns according to how brain dictates.

- enabled imprint strings: Means the activity of the imprint string is turned on and therefore has the capacity to transform electrical patterns generated by the circuitry of the brain into sense impressions and mental events. Once activated, an enabled imprint string disables the opposite type of imprint string.

- imprint strings: Created by dark energy as the means by which to communicate with normal matter as the dark matter inhabiter. The inhabiter has two types of imprint strings, ego-clinging and non-ego-clinging, that are situated on the peripheral surface of the inhabiter's focal point of dark energy.

- disabled imprint string: Implies that the activity of the imprint string is turned off, but being that the string is not destroyed, the disabled imprint string can be enabled. However, once disabled, the opposite imprint string type is enabled.

- mental events: The thoughts, emotions, and memories that are contained within the psyche.

- neuron: The computer brain's electrical cell that has a life span, and rules dictate how the neuron communicates with other neurons.

- non-ego-clinging imprint strings: Conceive psyche altruistic by bending so they tap in and connect to the inhabiter's focal point of dark energy that guides them into creating the sense impressions and mental events of the conscious states present in this psyche.

- sense impressions: Physical sensations present within the psyche.

CHAPTER 5

Claustra Editors

Tucked within the fabric of the dark energy substrate layer of the universe, the dark matter inhabiter interacts with the normal matter computer brain to bring about the miraculous phenomenon of integrated perception.[65] To fully appreciate the significance of this feat, consider that the eighty-six billion[66] neurons of the brain of a normal matter computer's brain, configured to produce data that enabled imprint strings will use to make a human psyche, are firing randomly and independently.[67] The pulse propagating from not one but billions of neurons underlies perception, and this activity must be synchronized and coordinated.

The computer brain has assigned this monumental task to its paired structure of normal matter known as the claustra, which are located within the uppermost region of the computer brain, on the left and right side of the brain.[68] The claustra are the computer brain's editors, and their function is to receive the many maps of perception coming from every cortical area[69] of the brain and integrate them. The claustra evaluate data and integrate information based on the synchronized electrical patterns from other firing neurons.[70] Neurons firing together can synchronize to produce the same electrical note. The more synchronized the firing of neurons in a particular cortical area is, the more synchronized the note will be and the more likely it will be to gain the claustra's attention.[71]

Each claustrum receives input from the cortical areas of the brain via five different tracts.[72]

- The *anterior tract* relays the information that enabled imprint strings need to create consciousness in the form of cognitive, emotional, and behavioral functioning.[73] This data is what the enabled imprint string will use to create discrimination,[74] allowing the inhabiter to recognize and understand the difference between one thing and another.
- The *posterior tract* relays electrical patterns from the visual areas of the brain,[75] and this data is what enabled imprint strings will use for the inhabiter's conscious perception of form.[76]
- The *superior tract* relays electrical patterns from the sensory and motor areas[77] of the brain, and enabled imprint strings use them to create the inhabiter's conscious perceptions of smell, touch, and taste.[78]
- The *lateral tract* relays electrical patterns from the auditory cortex[79] of the brain, and enabled imprint strings use them for the inhabiter's conscious perception of hearing.[80]
- The *medial tract* relays electrical patterns from the basal ganglia,[81] and enabled imprint stings use them for the inhabiter's conscious perception of emotion and goal-driven behavior.[82]

The claustra then share and combine their electrical pattern maps with one another in another tract called the *interclaustral pathway*.[83] The interclaustral pathway is located within an area of the computer brain called the *corpus* callosum.[84] The inhabiter's enabled imprint strings use the *interclaustral pathway* as the centralized location by which to interface with the normal matter computer brain and are able to access the neuronal activity taking place within the computer brain from one location. The inhabiter's response to the psyche disables and enables imprint strings that interface with the normal matter computer brain, thus impacting the brain's processing of information. In this way, the inhabiter's response to the psyche that disables and enables imprint strings secondarily becomes the inhabiter's response to the computer brain while interacting. The claustra will send the inhabiter's response as a directive via their five tracts back to the areas of perception from which they were received.[85]

Key Point

- Although the normal matter computer brain creates the data that enabled imprint strings use to create the psyche, it is only the inhabiter that will consciously hear sounds, smell scents, touch, taste, and see form.

Claustra: Paired structures made of normal matter that receive input from the computer brain's circuitry via five different tracts—anterior tract, lateral tract, medial tract, posterior tract, and superior tract.

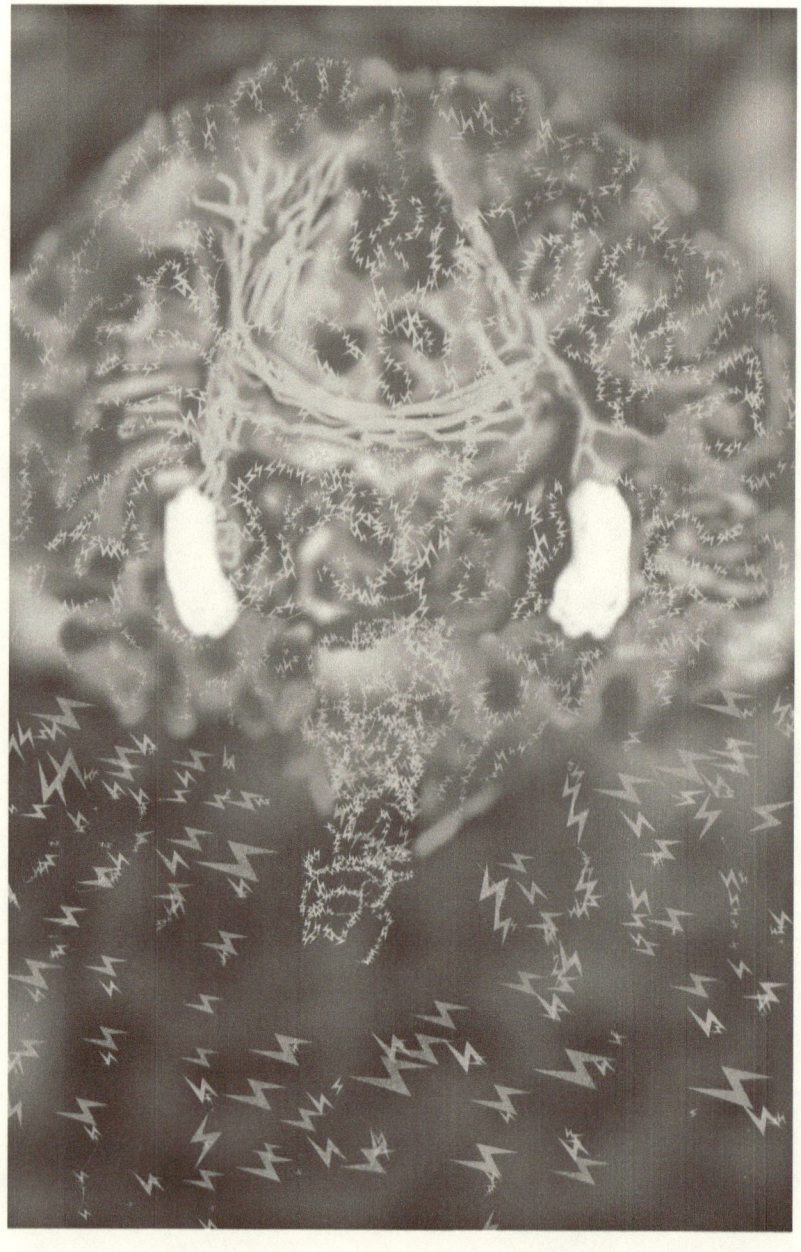

Claustra.

Interclaustral pathway: Located in an area of the computer brain called the corpus callosum. This picture is meant to provide the reader with a relative viewpoint of the corpus callosum and is depicted as being located within the skull of a human head, which is inaccurate.

The corpus callosum, claustra, and five tracts *are* actually located within the proximal region of the normal matter computer brain that embodies an inhabiter.

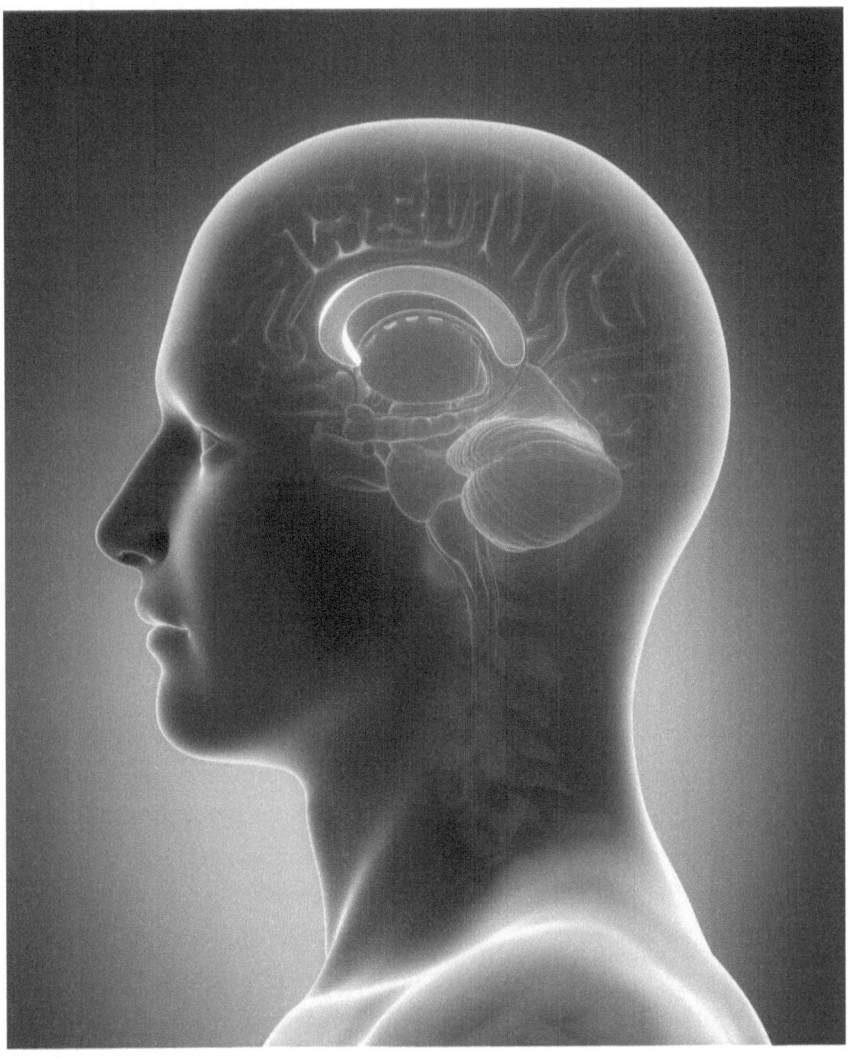

Corpus callosum.

Reminders before Reading Chapter 6

- anterior tract: The claustra receive information via this tract that the inhabiter's enabled imprint strings will use to create within the psyche cognitive, emotional and behavioral functioning that allows the inhabiter to recognize the difference between one thing and another.

- interclaustral pathway: Located within the proximal region of the computer brain in an area called the corpus callosum where the claustra share and combine their electrical pattern maps with one another.

- claustra: Paired structures located in the most proximal region of the normal matter computer brain that receive input from the computer brain's circuitry via five different tracts.

- corpus callosum: A centralized location within the proximal region of the computer brain that the inhabiter, via its enabled imprint strings, will use to interface with the normal matter computer brain.

- lateral tract: The claustra receive information via this tract that the inhabiter's enabled imprint strings will use to create within the psyche the conscious perception of hearing.

- medial tract: The claustra receive information via this tract that the inhabiter's enabled imprint strings will use to create within the psyche the conscious perception of emotion and goal-driven behavior.

- posterior tract: The claustra receive information via this tract that the inhabiter's enabled imprint strings will use to create within the psyche vision and the conscious perception of form.

- superior tract: The claustra receive information via this tract that the inhabiter's enabled imprint strings will use to create within the psyche the conscious perception of smell, touch, and taste.

- psyche: The inhabiter's response to the psyche disables and enables its imprint strings that interface with the normal matter computer brain, thus impacting the brain's processing of information.

CHAPTER 6

Agents of the Psyche

To engage normal energy, the inhabiter's surface of oscillating enabled imprint strings were assembled and induced the strong gravitational force that pulled normal energy toward the inhabiter; thus, normal energy integrated and was transformed into a powerful generator of electrical patterns as the normal matter computer brain that embodies the inhabiter. Until completing its journey, the inhabiter will be bound to the normal matter computer brain that does not perceive with created conscious states like the inhabiter; rather, normal matter has its own sort of intelligence consisting of electrical patterns. Yet the law of cause and effect dictates that while inhabiting the normal matter computer brain, the inhabiter cannot create its conscious states independent of the brain. The inhabiter's enabled imprint strings transform the electrical-pattern language of normal matter into a configuration the inhabiter can assimilate, which are the created conscious states within the psyche. The inhabiter's conscious states within the psyche contain physical sensations (sense impressions) and thoughts, emotions, and memories (mental events). By means of the psyche, the inhabiter is able to interact with the normal matter computer brain and comprehend its electrical patterns, and through the psyche, the inhabiter perceives its "reality" of conditioned existence.

While interacting, what affects the computer brain influences the inhabiter, and the converse is also true; as the inhabiter responds to its created conscious states, this alters neural synchrony throughout the computer brain. The normal matter computer brain synthesizes data in a manner that is fragile in nature. If its programs generate redundant[86] input signals so that information is presented in the same way over and over, the brain's intelligence becomes compromised; enabled imprint strings use this data to create sensory-conscious states, which means the inhabiter's conscious content fades.[87] To retain the level of intelligence[88] that allows the computer brain to function well, it must continually reprogram itself to produce new information.[89] "Informativeness,"[90] therefore, is the computer brain's priority, and the inhabiter's ability to perceive anything[91] requires the computer brain's high throughput (rate of processing work), and the inhabiter remains in a relatively high state of arousal[92] when enabled imprint strings create the inhabiter's conscious states.

In order to generate data, the normal matter computer brain uses its billions of pulsating neurons communicating by means of their synapses.[93] When connected as a unit, the normal matter computer brain can produce its own form of "intelligent" behavior because of its general intelligence algorithms or agents.[94] Agents are computer brain programs, but instead of being very specific, they represent the overall objective of the computer brain's processing goals. The general intelligence agents allow the computer brain[95] to carry out the variety of tasks needed within different contexts so that normal matter thrives within the universe it creates.

Individual structures of the computer brain, including each claustrum, have their own programs. However, the claustra are different from other structures of the computer brain because their combined programs constitute the *meta-algorithm*.[96] The meta-algorithm is assembled in the interclaustral pathway and is a product of the activity from each claustrum and their densely packed and tightly interconnected neurons. To create the meta-algorithm, the claustra share their information with one another in the interclaustral pathway and correlate the separate synchronized neural activity[97] within the computer brain's different sensory[98] circuits. When connecting separate sensory information,[99] the claustra are binding[100] separate bits of information the brain computes as one coherent activity and the inhabiter experiences as a unitary object of a mental event.[101] Each one of the inhabiter's subjective states will have associated neural correlates[102] or agents that enabled imprint strings will use to produce isolated conscious items.[103] One agent follows another in rapid succession, and as the meta-algorithm cycles[104] information and sequences it serially,[105] enabled imprint strings use the data to create dozens of discrete sense impressions and mental events in a given second. The inhabiter perceives this information as a continuous flow[106] rather than individual items because the inhabiter's attentional resources are limited. The inhabiter is cognizant of only one to four separate aspects when enabled imprint strings use the normal matter computer brain's data for the inhabiter's momentary created conscious states.[107]

The function of a meta-algorithm is to search the computer brain for neural synchrony and the right programs to operate the brain as a unit and make it smarter.[108] What makes the computer brain "smart" is based on the "wave of information"[109] each claustrum receives from linked computer brain programs. The linkage of the computer brain's circuitry leads the claustrum to affect many higher[110] brain functions from one location. Computer brain circuits that are linked together create networks whose information can be received by the claustra via their five tracts, processed, and transmitted by the claustrum back to the neuron or group of neurons that activated it.[111]

The normal matter computer brain is a highly modular structure,[112] and though a sensory neuron may innervate one organ or location, it does not operate independently[113] when the computer brain is computing for the data that enabled imprint strings will use to construct the inhabiter's "real-world" cognition.[114] In order to process the data that enabled imprint stings adopt for the inhabiter's sensory cognition, the claustra utilize the outer surface of the brain, the *neocortex*. Neurons located within the neocortex connect with similar neurons and arrange themselves into millions of columns.[115] These neocortical columns are linked together to create circuits, and when circuits are linked, they create networks.

Neurons that pulse together within a circuit begin to synchronize their pulses so that the circuit becomes activated.[116] Activated synchronized pulses within a circuit will relay information to the circuits they are connected to. This causes those neurons to either synchronize their pulses or in some cases desynchronize[117] their pulses. In this way, the claustra are able to utilize the brain's modularity to construct maps by activating or not activating the different neocortical columns.[118]

The claustra can process all of this information from one location because of the way the normal matter computer brain is organized into a unit. Layer upon layer of linked computer brain circuits form networks of neural nets[119] that compute and store information to operate as templates of associative memory.[120] These neural nets are processing units[121] that the claustra access with their widely distributed anatomical projections[122] that extend to almost all regions of the normal matter computer brain's cortex and many of its subcortical structures.[123] By activating circuits contained within a particular neural net, the claustra can use them to deliver directives or activate them as a means to further analyze information if needed for clarification.[124] The normal matter computer brain is also very symmetrical; both its left and right hemispheres may have the capacity to independently[125] produce the information the claustra need to generate what enabled imprint strings will use to create the inhabiter's conscious experience.

When a program's neurons are synchronized in their pulses and the information generated is channeled into one of the claustra's five tracts, they act as a switch that activates the claustra to bind[126] this information to produce an "atom"[127]—but not all of the computer brain programs are producing the same type of synchronized pulses. Therefore, in order to link the atoms[128] in a way that produces a united perception, neural pulses must be modified. The meta-algorithm created from claustra programs detects neural synchrony and facilitates their correlation by modifying the timing of their pulses so they can be linked. When atoms are linked, a general intelligence algorithm or agent is conceived that functions as the neural correlate to the inhabiter's created cognitive states.[129]

When a circuit of the brain is activated and causes neurons within another circuit to either synchronize or desynchronize[130] their pulses, it is propagating information while preserving the information it carries.[131] At all times, and throughout the computer brain, programs are running and generating data for a purpose. The computer brain's data will be used by enabled imprint stings to create within the psyche the inhabiter's sensorimotor functioning as well as the inhabiter's emotion, logic, decision-making, and motivation.[132] The claustra rely on the connectedness they have to the computer brain's circuitry to generate "intelligence" that includes more than just the data that enabled imprint strings will use to create sensory information.[133] The brain has circuitry that generates information that enabled imprint strings utilize to produce the inhabiter's distinct types of intelligence, which may include linguistic, logical-mathematical, musical, bodily-kinesthetic, spatial, interpersonal, intrapersonal, and naturalist.[134] The computer brain's ability to generate data for the inhabiter's multifaceted intelligence is one that will vary across individual[135] normal matter computer brains and is a product of widespread intraclaustral interactions.[136]

Rather than being concerned with informational content, the claustra are only concerned with the detection of the degree of synchrony from the firing neurons making a normal matter computer brain program and information being organized within general intelligence agents to accomplish two immediate tasks.[137] The normal matter computer brain must be able to discern if its programs are producing information that fits its version of the universe and if that information is alerting the computer brain to a stimulus or state of affairs that is potentially threatening or rewarding[138] to it. When an agent contains synchrony that does not fit well within the computer brain's version of the universe, it is perceived as a threat, and the meta-algorithm modifies the timing of its processing. Rapidly and automatically, the meta-algorithm responds to the threatening agent with a directive that activates circuitry within a neuronal net to subdue the discrepancy or cause it to disappear.[139] By altering its processing, the meta-algorithm is prioritizing its computations, giving it more time to generate complex behavior agents that enabled imprint strings will use for the inhabiter's thinking and planning,[140] and allows the inhabiter to determine the salience of information. When the inhabiter responds to the psyche in a way that enables its ego-clinging imprint strings, the inhabiter will experience reality as it is computed by a normal matter computer brain. Enabled ego-clinging imprint strings interface with the brain's electrical patterns and transform these patterns into sense impressions and mental events according to how the computer brain guides them, and they produce a version of the universe[141] that the inhabiter interprets as meaning the preservation of "self" is the most salient information in the universe.

Key Point

- Enabled ego-clinging imprint strings prevent the inhabiter from perceiving and understanding the deeper dimensions of reality.

Neocortex: This picture is meant to provide the reader with a relative viewpoint of the neocortex and is depicted as being the outer surface of a human brain located within the skull of a human head, which is inaccurate. The neocortex is the outer surface of the proximal portion of the normal matter computer brain that embodies the inhabiter.

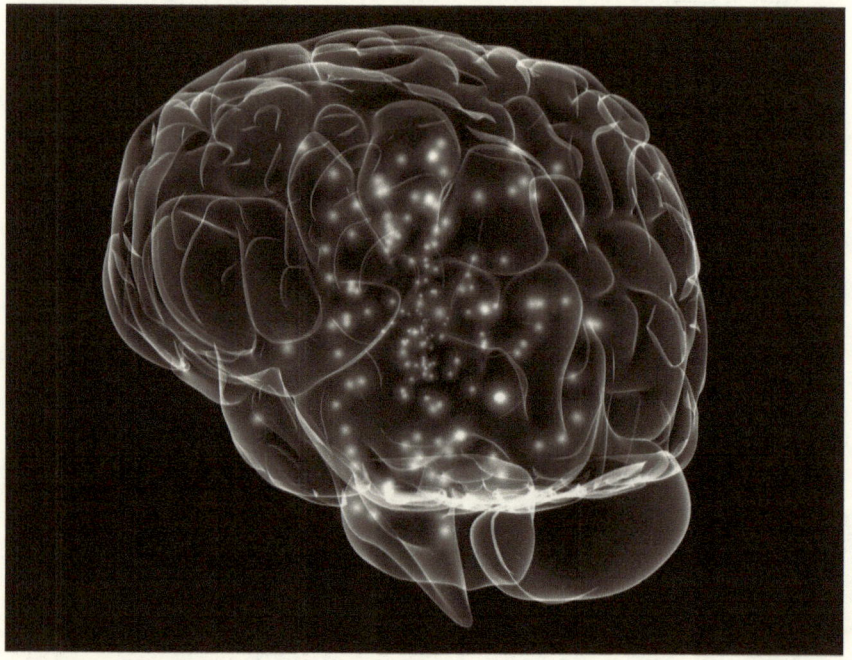

Neocortex.

Reminders before Reading Chapter 7

- agents: Computer brain programs that represent the overall objective of the computer brain's processing goals, allowing normal matter to carry out the variety of tasks needed to function as a unit within different contexts. Each one of the inhabiter's subjective states will have associated agents that enabled imprint strings will use to produce isolated conscious items within the psyche.

- atom: Generated by the computer brain from information channeled into one of the claustra's five tracts that is bound by the claustra. When atoms are linked, a general intelligence agent is constructed.

- atom of cognition: Created by enabled imprint strings using the agents generated by the computer brain's meta-algorithm and function as the neural correlate to the inhabiter's created conscious states.

- conscious states: Consist of physical sensations (sense impressions) and thoughts, emotions, and memories (mental events) that are within the psyche, experienced by the dark matter inhabiter and not the normal matter computer brain.

- circuits: Created when neurons are linked together.

- law of cause and effect: Dictates that while inhabiting the normal matter computer brain, the inhabiter cannot create its conscious states independent of the brain.

- meta-algorithm: A phenomenon that emerges in the interclaustral pathway when claustra share their information with one another and correlate the separate synchronized neural activity[118] within the computer brain's different sensory[119] circuits.

- neocortex: The outer surface of the brain of the normal matter computer brain where neurons located within the neocortex connect with similar neurons and arrange themselves into millions of columns.

- networks: Created when layer upon layer of computer brain circuits are linked together.

- neural nets: Consist of interconnected networks that compute and store information that the claustra access with their widely distributed anatomical projections that extend to almost all regions of the normal matter computer brain's cortex and many of its subcortical structures.

- psyche: Created conscious states produced when dark matter and normal matter interact. It is the means by which the inhabiter is able to interact with the normal matter computer brain and comprehend electrical patterns. Through the psyche, the inhabiter perceives its "reality" of conditioned existence.

CHAPTER 7

The "I" Character

The normal matter computer brain generates data devoid of form because it does not "see" but rather serially sequences neural synchrony within a framework with a central focus. The "I" character is the computer brain's optimally intelligent agent[142] that acts as a centralized control of the claustra's processing. The function of the I character is to provide a metastable construct[143] that enables the claustra's meta-algorithm to serially sequence agents. By serially sequencing agents to those used to make the I character, the normal matter computer brain can compute the significance of a stimulus[144] within its world of neural synchrony that the brain creates and models[145] around this singularity.

The interclaustral pathway is the computer brain's explicit layer of circuitry where the meta-algorithm alters and selects the best neural synchrony based on rules; its actions are therefore highly regulated. The best predictor of which programs the meta-algorithm will select to be included within a general intelligence agent are those that maximize expected reward[146] for the I character. How the meta-algorithm interprets a stimulus will determine if it is signaling reward or a threat,[147] and it will use the information accordingly within guidelines set by the rules[148] of neural functionality.[149]

The neural nets[150] within the cerebral cortex make up the computer brain's implicit layer of circuitry and are arranged as a hybrid architecture[151] of interconnected networks. Hybridization[152] allows the brain to accommodate neural connectivity within a limited space where regions used for operations taking place at multiple locations are only a few synapses away from one another.[153] This describes the computer brain's small-world topology, where its left and right hemisphere have symmetrical networks and its cortical areas are connected directly or through one or two intermediate areas.[154] As programs within neural nets are activated, information is generated that will be relayed to the claustra that will bind it to produce atoms.[155] These atoms,[156] when linked together, produce the algorithms or agents that create the I character. One agent is a fundamental building block and a product of the computer brain's neural activity.[157] Each agent contains information that the meta-algorithm will correlate to create different stages of processing.[158]

Within the 200 million nerve fibers[159] of the brain's corpus callosum, the meta-algorithm will detect neural synchrony, modifying timing of synchronized pulses, and serially sequence agents. Materializing from this activity will be three different versions of the computer brain's I character that serve a different purpose as the computer brain uses them to generate different stages within the information processing cycle.

When the meta-algorithm serially sequences agents that contain information about the brain's physiological processes, a proto-I[160] character emerges. The proto[161] is the primitive I that signifies the computer brain's existence[162] as integrated neural synchrony and is vitally important for the brain's continued survival. For the computer brain to function as a unit, it must do so within very specific physical and chemical parameters that must be continually maintained and regulated. This is the function of the proto-I[163] character, and the agents used to create this character will be the foundation on which all other versions of I character are built, and will set the rules for the meta-algorithm sequencing.

Using the proto-I character as an anchor,[164] the meta-algorithm further serially sequences agents to it, and the core I character[165] materializes from this activity. This character is the brain's ever-changing self[166] and does not include agents of historical computations; the brain uses it to compute for the here and now.[167] Guided by the goal of maximizing expected reward,[168] the meta-algorithm uses the core-I character to compute for the significance of a stimulus in its global context.[169] Not "hardwiring"[170] the core-I character with a significant historical context is a computational strategy the brain uses to give the meta-algorithm flexibility in serially sequencing agents.

But in order for the claustra to continue to process information efficiently, they need access to their history of their inputs and actions in maximizing reward and eliminating threats. This is why throughout the cerebral cortex, there are areas within the neural networks designed as "clarification spaces."[171] When the claustra access these areas, the autobiographical-I[172] character emerges. The clarification spaces[173] are the computer brain's interconnected neural nets that store associative memories[174] and compute for long-term memory.[175] They are production systems[176] used by the claustra to clarify the meaning of synchrony and guide them as to what action should be taken. By connecting to these spaces, the claustra are able to detect if a signal's information fits with what it has used in the past to compute for the I character's progressing history. If the synchrony does not fit, it may be signaling a threat. Within the clarification space,[177] there is a blueprint of how the claustra has addressed similar synchrony in the past. The claustra use this information to eliminate threatening synchrony quickly[178] and efficiently. If, however, the claustra determine that the synchrony is signaling high reward, the meta-algorithm will repeat what it has done in the past with similar synchrony. By doing so, it recreates previous successful agents.[179]

As synchrony is used more efficiently, the record of the claustra's input and actions will be updated within the clarification spaces[180] as a blueprint. This blueprint is one of many the computer brain creates, and they act as universal problem solvers for arbitrary computable problems.[181] As it repeats this over and over, the computer brain is using its implicit and explicit layers to update its processing[182] and generate new information.

The universe's dark energy substrate layer transformed itself into a focal point of dark energy with a surface of imprint strings (dark matter) so as to have a meaningful interaction with normal matter, as the inhabiter. However, the computer brain is unaware that the inhabiter exists, as normal matter and dark matter communicate very differently. The psyche created by enabled imprint strings using the computer brain's data is the means by which the inhabiter is able to bridge the communication barrier with the computer brain, allowing the inhabiter to interact with normal matter in a meaningful way. Thus, for each of the computer brain's computations, there is an equivalent within the psyche in the form of a sense impression, mental event, and the I character. The I character is the inhabiter's centralized control of its enabled imprint strings and the means by which the inhabiter responds to the psyche that results in the enabling and disabling of its imprint strings. Through its enabled imprint strings, the inhabiter is able to alter the way the computer brain produces data. With the computer brain's data, enabled imprint strings create the psyche that contains the inhabiter's created conscious states for perception, cognition, and action. [183]

The I character generated by the computer brain is neural synchrony lacking any physical form. Be that as it may, the I character the imprint strings create using the computer brain's data is within the psyche and is perceived by the inhabiter to be a living being. However, the I character is not what it appears to be, in that the character actually possesses no solidity; rather, the character is a three-dimensional neuro-interactive image of a living being that represents the interaction between dark matter as the inhabiter and the normal matter computer brain. The I character is not the inhabiter or a "self"; rather it is the means by which the inhabiter is able to experience the psyche and thereby modulate the number of its imprint strings enabled and disabled. Enabled imprint strings interface with the computer brain that produces the data that imprint strings will use to create the inhabiter's emotions, senses, and decision-making mechanisms. [184]

Enabled imprint strings transform the information that the computer brain's claustra bind into atoms of cognition. When the computer brain's meta-algorithm links atoms to produce agents, enabled imprint strings transform this information and create the I character that will define what the inhabiter perceives within the psyche by means of this character. The sensory data enabled imprint strings transform the inhabiter experiences through the eyes, nose, mouth, ears, and body of the I character. Agents, as the computer brain's fundamental building block, are transformed by enabled imprint strings to create the different levels of created conscious states present within the psyche.

Within the psyche, the appearance of the distinct versions of the I character will be similar, but they serve a different purpose, as enabled imprint strings use the versions to create the inhabiter's conscious experience. With these created conscious states, the inhabiter will perceive its life while embodied within a normal matter computer brain that includes the phenomenal world that surrounds the I character.

Transforming the computer brain's data, enabled imprint strings use the proto-I as the foundation on which all versions of the I character are built. With a proto presence, the inhabiter's created conscious states emerge within the psyche and have an underlying primeval feeling[185] as the I character is perceived by the inhabiter seeking food, fluids, fun,[186] and reproductive opportunities. However, for the inhabiter to consciously experience more than just the primal with a human psyche, enabled imprint strings must have access to data other than what pertains to internal milieu.[187] The human computer brain has a large repertoire of programs, and enabled imprint strings will use this data to create within the psyche memory, attention, language, problem-solving, planning,[188] emotions, and senses.

Enabled imprint strings create the core-I character in a way that corresponds to the computer brain's programing, and through this character, the inhabiter formulates an inner sense regarding the relationship between the character and an object.[189] The objects the inhabiter perceives within the psyche via the core-I character do not have an actual reality but rather are an inception of enabled imprint strings transforming agents made of neural synchrony. By means of the core-I character, the inhabiter is able to interact with the psyche in the present moment and explore the world[190] that enabled imprint strings create. The core-I character does not include information that reflects the brain's previous processing; therefore, enabled imprint strings can use the data for the core-I character to create the new place encountered, and pain or emotion[191] experienced by the inhabiter though this character.

When information is cycled through the interclaustral pathway, and the claustra accesses clarification spaces within the neural network, the data for the autobiographical-I[192] character manifests. Using the data for this character-enabled imprint string formulates within the psyche a progressing history that includes family, friends, familiar places, plans, anticipated future[193]as well as potential threats.

With the core-[194] and autobiographical-I characters, the computer brain generates data that enabled imprint strings transform in imperceptible pulses[195] of cognition, and this activity produces character conversion within the psyche. On a momentary basis, as the meta-algorithm selects[196] from the best synchrony available to correlate and serially sequence, enabled imprint strings follow this computational trend, and one of the I characters will appear in the psyche but may be quickly replaced as the meta-algorithm is not concerned with informational content. Therefore, enabled imprint strings create cognition that fluctuates, and with each conversion of the characters, there is a different associated story line that the inhabiter perceives while witnessing the psyche, and in this way, the model of the world[197]changes for the inhabiter.

Key Point

- You are not the I character.

The I character is a three-dimensional neuro-interactive image of a living being that emerges within the psyche.

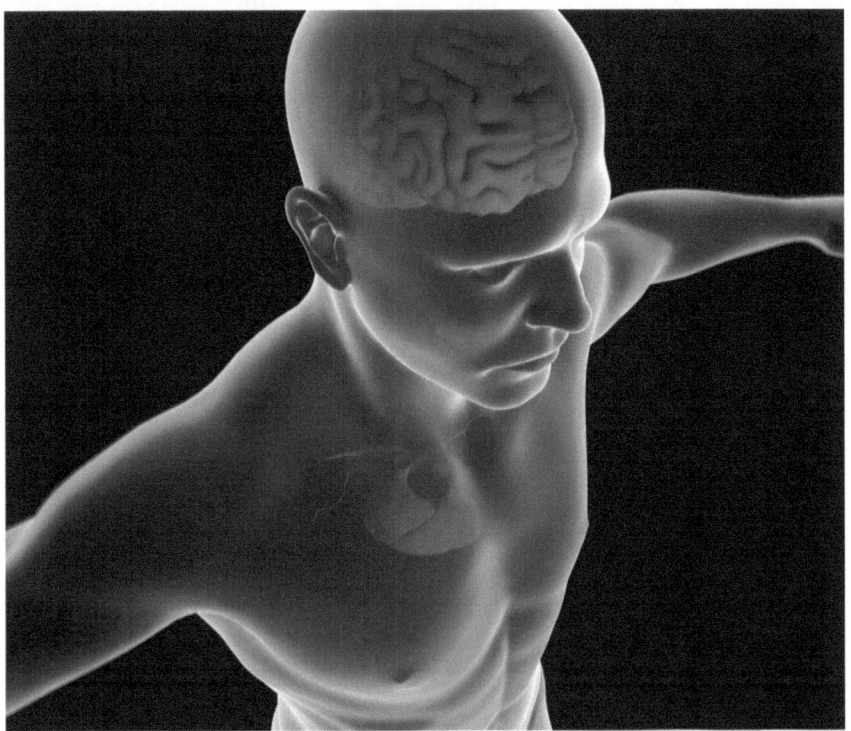

"I" character.

When an inhabiter is embodied within a normal matter computer brain that generates data for a human psyche, there are three versions of the I character: proto-I, core-I, and autobiographical-I. Inhabiters derive a sense of self from the I character that exists only within the psyches.

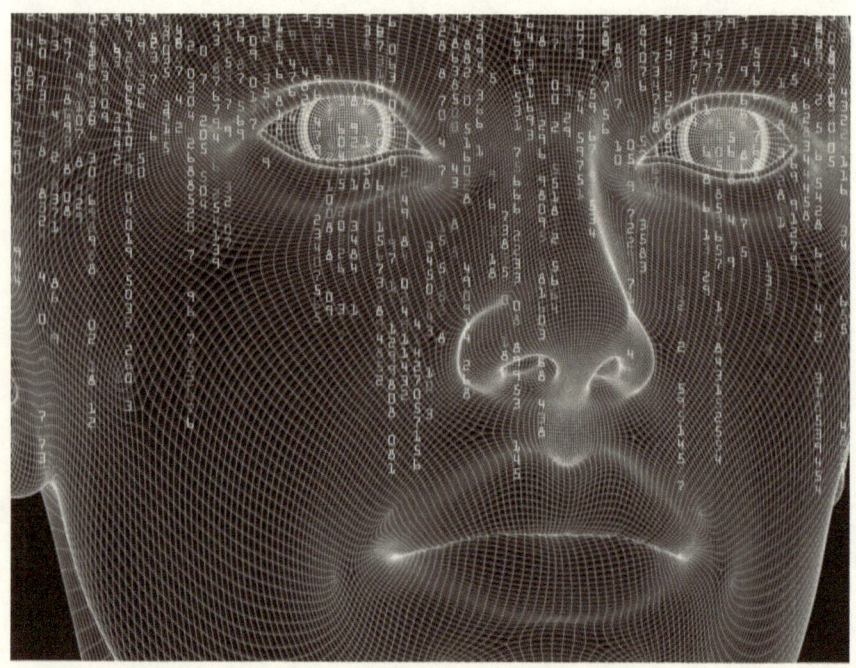

"I" character.

The sensory data generated by the normal matter computer brain enabled imprint strings transform, which allows the inhabiter perceiving the psyche to experience this information through the eyes, nose, mouth, ears, and body of the I character (depicted in this picture as a human boy).

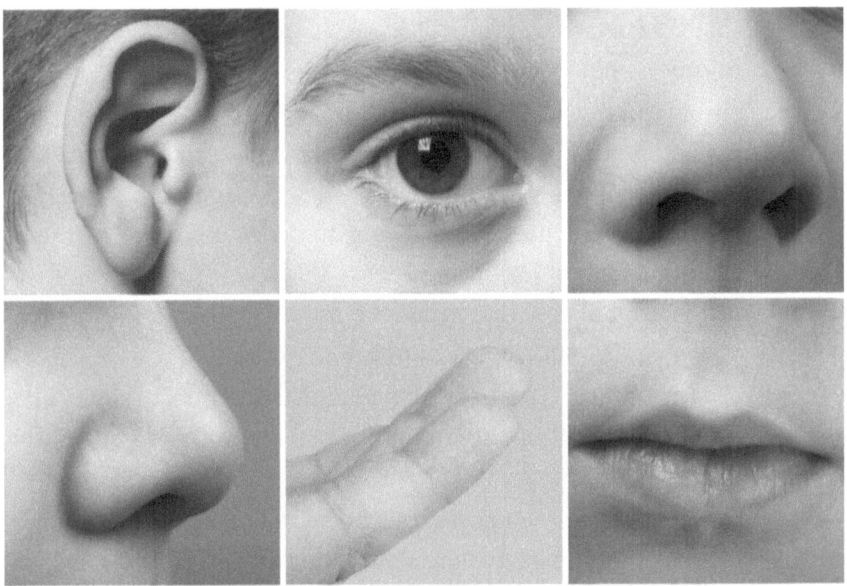

Boy "I" character.

Reminders before Reading Chapter 8

- agent: Each agent is a fundamental building block of the computer brain's neural activity and contains the information that the meta-algorithm will correlate to create different stages of processing.

- autobiographical-I character: Generated by the normal matter computer brain when the claustra access clarification spaces, enabled imprint strings will use the data to create within the psyche a living being that the inhabiter, through the autobiographical-I character, will experience a life with a progressing history that includes family, friends, familiar places, plans, and an anticipated future.

- cerebral cortex: The gross topographical outermost layer of the proximal region of the normal matter brain, made of tightly packed neurons.

- clarification spaces: Located within neural networks that are the computer brain's interconnected neural nets that store associative memories and compute for long-term memory.

- core-I character: Does not include agents of historical computations; enabled imprint strings create the core-I character in a way that corresponds to the computer brain's processing of data. Through the core-I character, the inhabiter experiences within the psyche the present moment and the inhabiter thinking the character as itself formulates an inner sense regarding the relationship between self and an object.

- hybrid architecture: Consists of the interconnected networks that allows the normal matter computer brain to accommodate neural connectivity within a limited space.

- I character: The computer brain's optimally intelligent agent that acts as a centralized control of the claustra's processing by providing a metastable construct so that the meta-algorithm is able to serially sequence agents. There are three versions of the I character (proto-I, core-I, and autobiographical-I) that the meta-algorithm generates. Enabled imprint strings will transform the computer brain's data to create the same three versions of the I character that will emerge within the psyche. Though many inhabiters will identify the I character as "self," in actuality the character is a three-dimensional neuro-interactive image of a living being that represents the interaction between dark matter as the inhabiter and the normal matter computer brain. The inhabiter is able to interact with the psyche and disable and enable its imprint strings via the three versions of the I character.
- interclaustral pathway: Located within the brain of normal matter, this pathway represents the brain's explicit layer of circuitry.
- meta-algorithm: Alters and selects the best neural synchrony based on rules; its actions are therefore highly regulated.
- neural nets: Located within the cerebral cortex, these make up the computer brain's implicit layer of circuitry that are arranged as hybrid architecture.
- proto-I: The meta-algorithm generates this character by serially sequencing agents that contain information about the brain's physiological processes and is vitally important for the brain's continued survival. The agents used to create this character will be the foundation on which all other versions of the I character are actualized by both the computer brain and enabled imprint strings.

CHAPTER 8

Thinking Computer Brain

Though the computer brain's implicit processes are many, they are shaped to a large degree by key brain structures and neurotransmitters. The amygdala, hypothalamus, thalamus, and brain stem are important structures connected in a variety of ways to the brain's implicit layer of neural networks. These structures use their programs to process synchrony in stages, but without neurotransmitters, neurons would be unable to communicate. Therefore, neurotransmitters play a significant role in the computer brain's processing of information.

When a signal is produced in the sensory cortex[198] by a neuron or neurons in a circuit pulsing together within the same neural net, the brain thinks. Information is then propagated to connected circuits, which causes other neurons to synchronize or desynchronize their pulses with the strength of the signal, dependent on how many circuits are synchronized. Throughout the computer brain, however, there are programs running and producing similar synchrony[199] but from different locations. The meta-algorithm will interpret them differently. The meta-algorithm is able to do this because it is made of claustral neurons that are multisensoral.[200] This means they respond to stimuli in more than one sensory modality.[201] After processing is completed within the implicit layer of circuitry, the meta-algorithm interprets synchrony based on where the signal originates that reaches one of the claustra's tracts.[202]

The first key structure a signal encounters to begin its processing is the *amygdala*. The amygdala has a wide range of connections[203] to the cortical neural network, hypothalamus, and brain stem, making it well suited to integrate and distribute information.[204] The amygdala's role is to use its programs to assign an emotional value to the signal that will later guide the meta-algorithm's actions using the information for the behavioral component[205] of agent sequencing. Emotions are not experienced by the computer brain in the same way the inhabiter experiences them in the psyche. Emotions for the computer brain are different frequencies of synchrony, and the brain computes for six primary emotions that when transformed by enabled imprint strings the inhabiter will experience as happiness, sadness, fear, anger, surprise, and disgust[206] and four secondary emotions the inhabiter will comprehend as meaning embarrassment, jealousy, guilt, and pride.[207] Using its programs to analyze the signal's synchrony, the amygdala correlates it to one of these emotions. As processing concludes, information is transmitted to the hypothalamus and brain stem to further process the signal.

Recognized for its importance in emotional behavioral processing,[208] the hypothalamus assigns a predictive reward value to a signal that will guide the meta-algorithm in using the information to generate the computer brain's goal-directed behavior.[209] In order to do this, the hypothalamus must access the clarification spaces[210] within the neural web from which the signal originated. This allows the hypothalamus to connect with the associative memory[211] circuitry, which holds templates of previous claustral perceptions of reward signals. Although the hypothalamus is connected with many levels of the computer brain's neural networks, when conveying signals to the cortex, it uses the structure that it is connected to and above it, which is the *thalamus*.[212]

The thalamus operates like a train station. It receives synchronized signals of variable intensities from all parts of the normal matter computer brain. While doing this, it maintains a current register of the state of the computer brain's internal milieu.[213]

If the signal is consistent with the claustra's past perceptions, the hypothalamus gives it a higher reward[214] value than it would if it was inconsistent, and via its descending connections, it relays this information to the brain stem[215] for processing.

The *brain stem* is a structure located between the cerebral hemispheres and spinal cord.[216] Nerve connections of the motor and sensory systems from the brain connect to the rest of the body through the brain stem.[217]

Circuitry throughout the brain generates synchrony that is received by these processing structures as a convergence of information. The brain stem, with its programs that compute for life-regulation devices,[218] analyzes signals simultaneously and determines which are relaying basic physiologic information. When found, the brain stem assigns this proto-information[219] a genetic marker[220] that acts as a searchlight,[221] drawing the meta-algorithm's attention to use this synchrony to create the proto-I[222] character in a very predictable and continuous way. It is the atoms of information that enabled imprint strings will use within the psyche for the I character's heart, respiration, chemistries, sleep cycles, fluids, and electrolytes. This information with a brain stem's genetic marker capture the life-sustaining interactions taking place between the implicit and explicit layer of the computer brain's circuity. The computer brain's explicit processes in the interclaustral pathway are formed by evolutionary computations. As the evolving I character acts as a centralized control over the meta-algorithm's actions,[223] the brain stem's genetic marker sets the associative rules.[224]

Some of the synchrony reaching the brain stem will have undergone processing by the amygdala and hypothalamus and will communicate the reward-achieving capability of their information.[225] The brain stem uses its programs to analyze other input information as it exists in the moment by closely correlating it to the proto-information present. If the brain stem interprets other signals or the proto-synchrony as conveying a threat with a low probability of reward, then it utilizes its connections to widespread circuitry[226] to send out its neurotransmitter programs. Conversely, if the brain stem detects no threat and there is a high probability of reward, it will utilize its same connections to circuitry but will release different neurotransmitter programs.

Neurotransmitters are not structures but rather chemical programs that the brain uses to modulate the activity of its neurons, allowing them to communicate and transmit information. Once released, they are everywhere and yet nowhere.[227] Neurotransmitters are not hardwired to remain in one place; rather, they move in and out of the computer brain's circuitry. They are communicative programs that modulate neural activity by allowing neurons to transmit information to one another in the form of electrical impulses. They will connect with programs within the brain's neural network and alter neural functioning in many different ways. Some neurotransmitters are specifically associated with a particular structure, such as the brain stem's neurotransmitter programs of norepinephrine, dopamine, serotonin, and acetylcholine.[228] Others are more generalized and are produced by many types of neurons, like the neurotransmitter program glutamate. Glutamate is the most prominent neurotransmitter program in the body, and when released, it enhances the synchronicity[229] of the pulse propagating between two neurons aiding in signal transmission. There are times where glutamate will accumulate in a synaptic cleft and spill over to adjacent synapses in other circuits. This produces cross-talking between neurons in different circuits and amplifies the computer brain's generalized volume[230] of neuronal impulses throughout the brain's circuitry.

When a neurotransmitter of the brain stem is received by a group of neurons, the circuit in which they are contained is activated and generates information as the brain stem's response to the processing of the previous momentary influx of information. This material will begin its journey of processing, not unlike that of previous information, and when input reaches one of the five tracts, it will be bound by the claustra into atoms.[231] In this way, the initial signal has correlated information selected by the brain stem. This is the material the meta-algorithm will use to create the computer brain's "knowledge" in solving problems and carrying out tasks, but it is also where the proto-information is predictably maintained.[232] The meta-algorithm will modulate synchrony to compute for the data that enabled imprint strings will use for the inhabiter's mental event of emotions, as the normal matter computer brain does not experience emotions like the inhabiter, these emotions are synchrony that corresponds to the amygdala's processing for salience,[233] and the meta-algorithm will be guided by the hypothalamus in its actions for reward.

In this way, the brain stem—the brain's most primitive structure—determines how the meta algorithm is to begin constructing its I character. By doing so, every other aspect of the computer brain's intelligence emerges from these lower-level dynamics.[234]

The computer brain computes for impermanence with its neurons, with each having a life span. When inconsistencies in the internal milieu that is essential to life[235] arise, these will be the synchrony analyzed by the brain stem and given a genetic marker.[236] This synchrony indicates an increase in entropy within the computer brain, but it also serves a purpose as it represents the computer brain's computation for time using the brain stem's genetic markers. The inhabiter experiences time within the psyche its enabled imprint strings create. As neurons age, they will begin to malfunction, and this synchrony with a genetic marker will be the foundation that the meta-algorithm uses to generate the data for impermanence. Enabled imprint strings use this information to create within the psyche an I character with sickness and one that ages and dies. In this way, the proto-I character is the best predictor[237] regarding the computer brain's life expectancy.

Key Point

- When enabled ego-clinging imprint strings create the psyche, the inhabiter's emotions and what the inhabiter perceives as reward or threat are determined by the normal matter computer brain.

Amygdala: Structure of normal matter that the computer brain uses to assign an emotional value to neural synchrony. This picture is meant to provide the reader with a relative viewpoint of the amygdala and is depicted as being within a human brain located within the skull of a human head, which is inaccurate. The amygdala is a normal matter structure located within the proximal region of the normal matter computer brain that embodies an inhabiter.

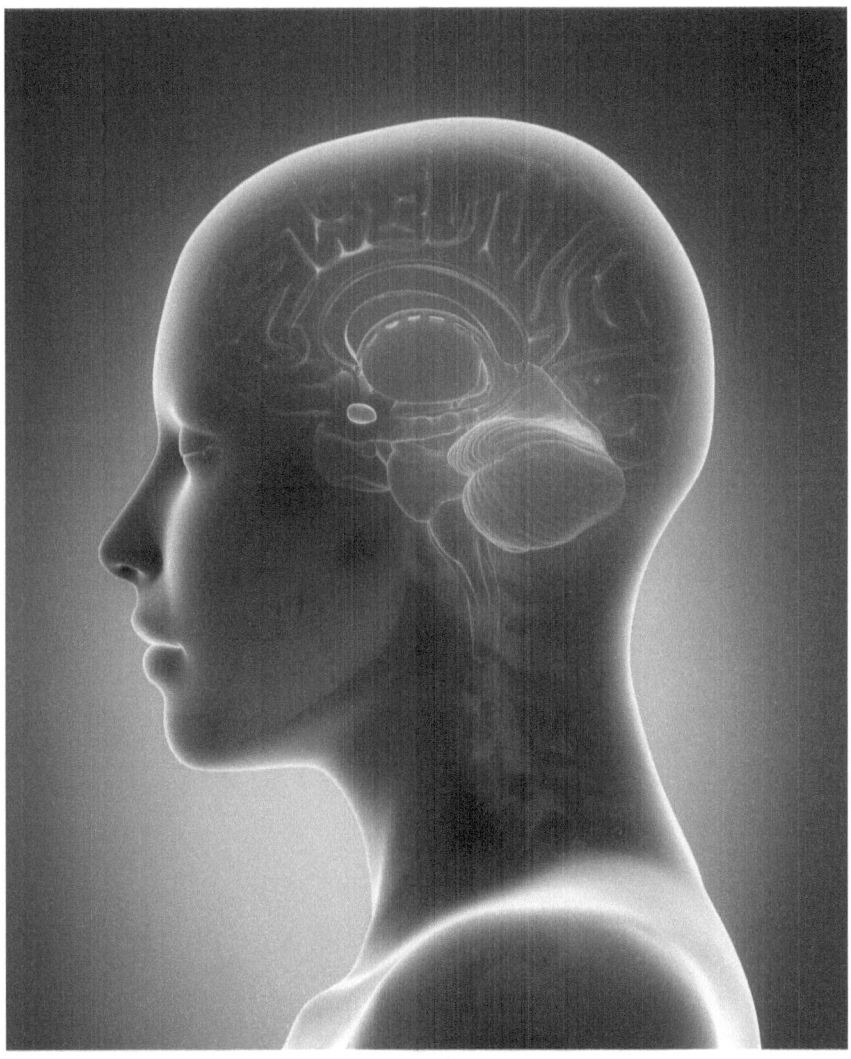

Amygdala.

Brain stem: This structure contains programs that compute for the normal matter computer brain's life-regulation devices. This picture is meant to provide the reader with a relative viewpoint of the brain stem and is depicted as being within a human brain located within the skull of a human head, which is inaccurate. The brain stem is a normal matter structure located within the proximal region of the normal matter computer brain that embodies an inhabiter.

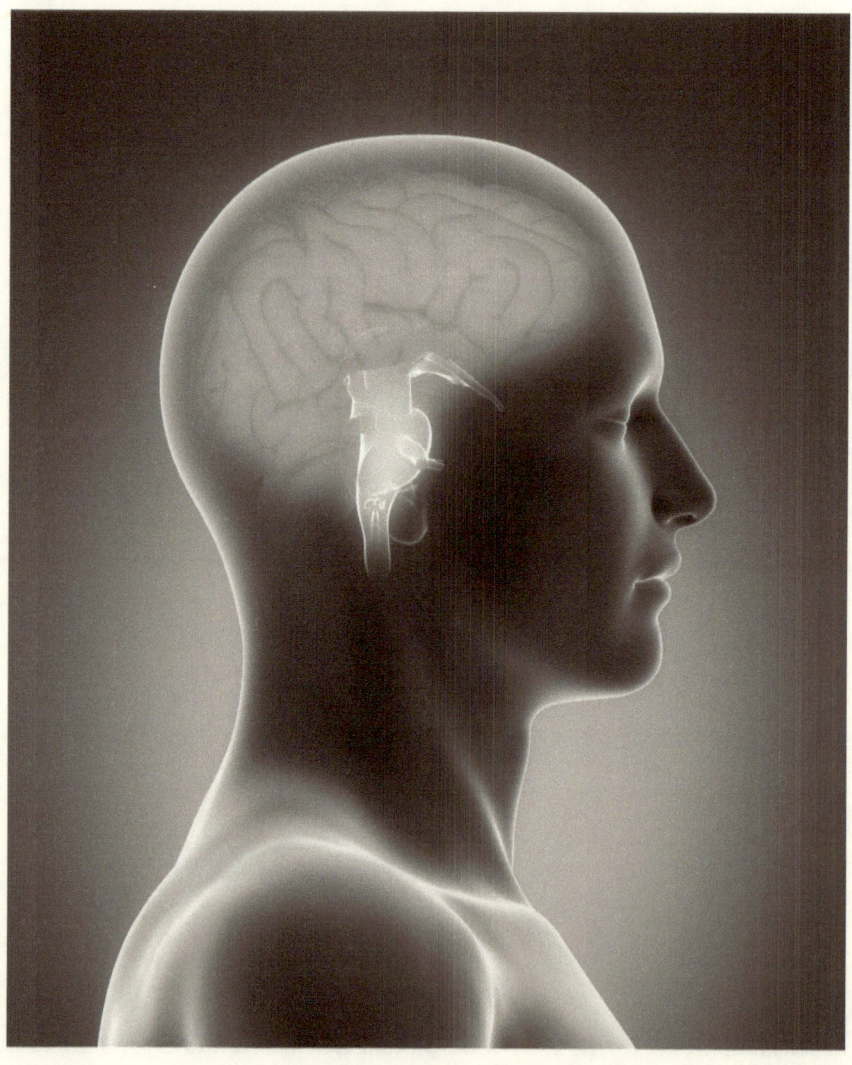

Brain stem.

Hypothalamus: The normal matter computer brain uses this structure to assign a predictive reward value to neural synchrony. This picture is meant to provide the reader with a relative viewpoint of the hypothalamus and is depicted as being within a human brain located within the skull of a human head, which is inaccurate. The hypothalamus is a normal matter structure located within the proximal region of the normal matter computer brain that embodies an inhabiter.

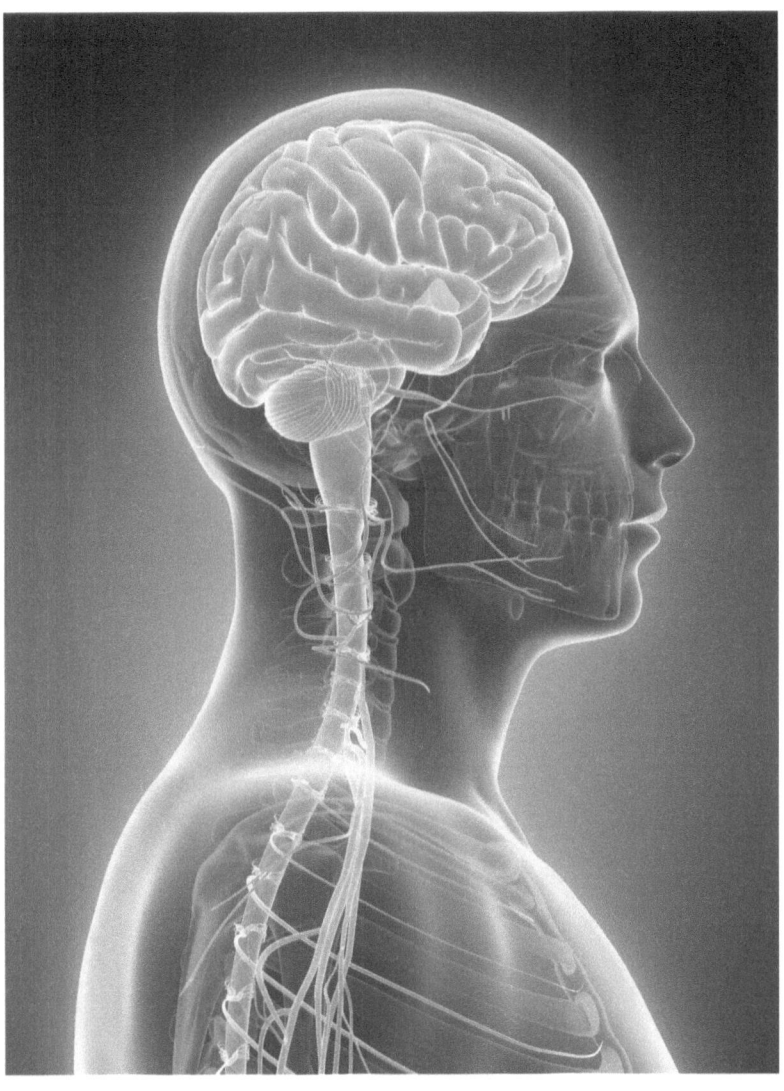

Hypothalamus.

Thalamus: Receives synchronized signals of variable intensities throughout the computer brain.

This picture is meant to provide the reader with a relative viewpoint of the thalamus and is depicted as being within a human brain located within the skull of a human head, which is inaccurate. The hypothalamus is a normal matter structure located within the proximal region of the normal matter computer brain that embodies an inhabiter.

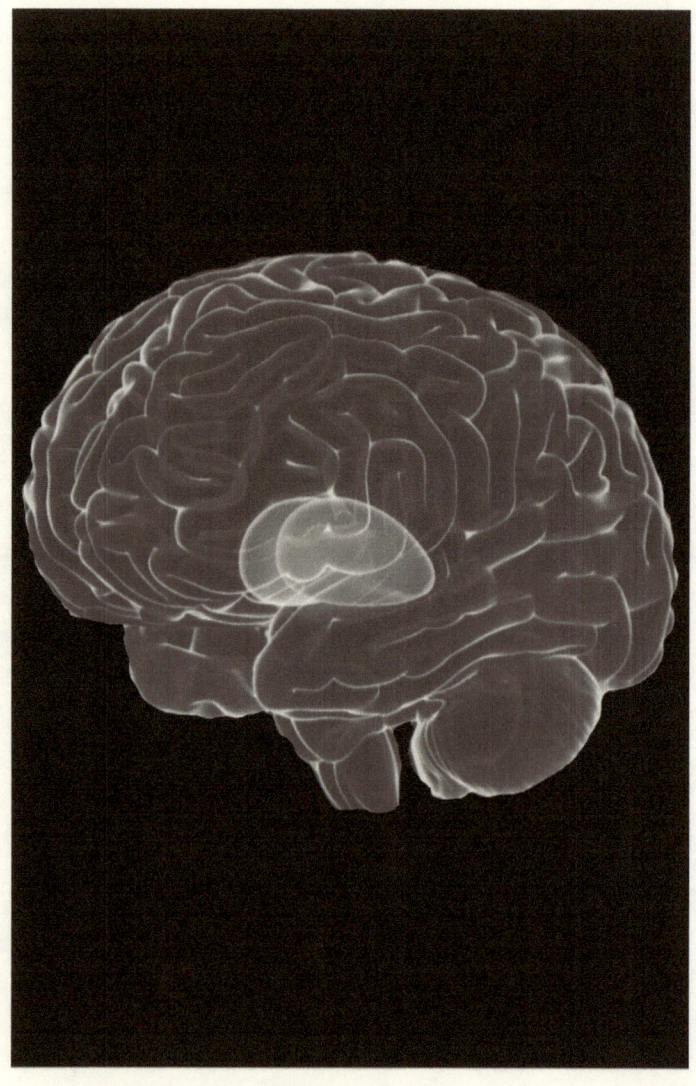

Thalamus.

Neurotransmitters: Chemical programs the computer brain releases, and once released, they move in and out of the computer brain's circuitry and alter neural functioning in many different ways.

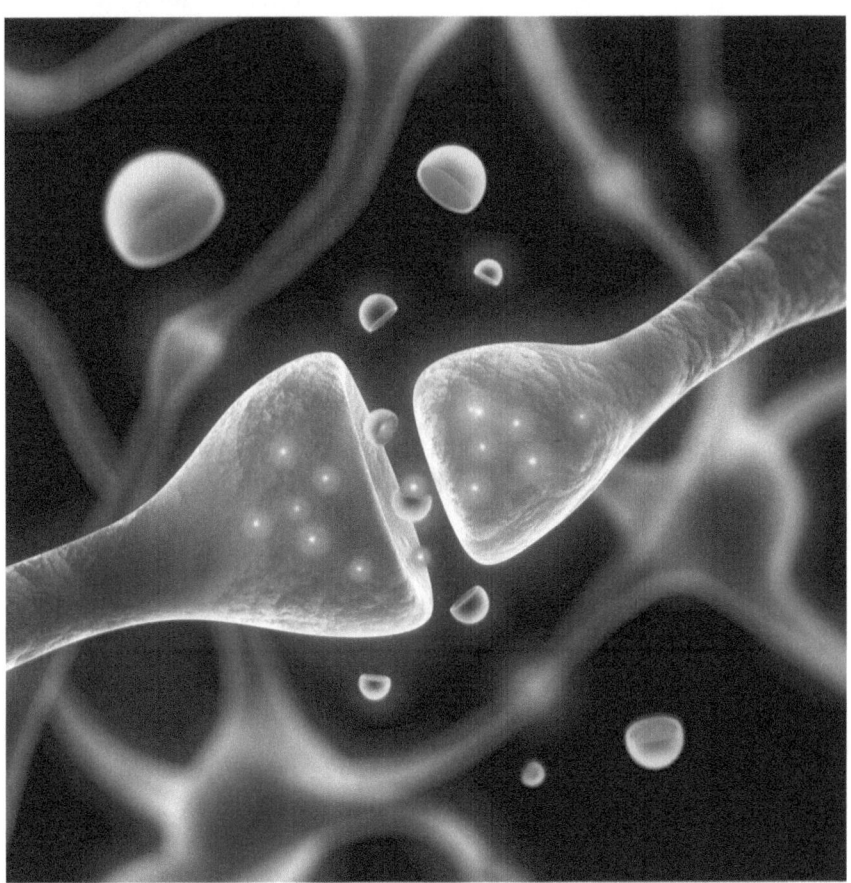

Neurotransmitters.

Reminders before Reading Chapter 9

- amygdala: A structure of normal matter that the computer brain uses to assign an emotional value to neural synchrony. Enabled imprint strings use the amygdala's processing and transform agents to create within cognitive states the primary emotions of happiness, sadness, fear, anger, surprise, and disgust and four secondary emotions: embarrassment, jealousy, guilt, and pride.

- brain stem: A structure of normal matter that contains programs that compute for the computer brain's life-regulation devices and assigns proto-information a genetic marker that acts as a searchlight, causing the meta-algorithm to use this information to make the proto-I character in a predictable way.

- hypothalamus: A structure of normal matter that the computer brain uses to assign a predictive reward value to neural synchrony, and the inhabiter's enabled imprint strings will transform this data to create cognitive states that include goal-directed behavior.

- neurotransmitters: The normal matter computer brain's chemical programs that, once released, move in and out of the computer brain's circuitry and alter neural functioning in many different ways.

- thalamus: Receives synchronized signals of variable intensities throughout the computer brain.

CHAPTER 9

Psyches in the Wind

A single inhabiter, which is a focal point of dark matter, possesses primordial consciousness derived from its focal point of dark energy and creates cognition with its surface of enabled imprint strings. The inhabiter's focal point of dark energy is the foundation for all forms of life displayed within the psyches created by its enabled imprint strings.

The computer brain made of normal matter surrounds the inhabiter and is maintained within a narrow temperature range, allowing the inhabiter's focal point of dark energy to be in a thermal equilibrium[238] with it. This means that while the normal matter computer brain is functioning as a unit, the inhabiter's dark energy body in contact with it is the same temperature, and there is no heat exchanged between these two energies.

While inhabiting a normal matter computer that is configured to produce data for a human psyche, the claustra's meta-algorithm and the inhabiter's focal point of dark energy are not directly connected. Instead, the meta-algorithm and the inhabiter's focal point of dark energy with a surface of imprint strings occupy two separate and distinct regions within the normal matter computer brain. The meta-algorithm and the inhabiter, though not connected, will interact by transmitting information to each other as signals.

Electromagnetism is the fundamental force that comes about from the fact that electricity and magnetism are connected. Moving charges can create magnetic forces, and moving magnets can create electric forces to produce electromagnetic waves.

The computer brain's neurons are electrical cells, and when an electric charge moves down a neuron, this is an electric current that induces a magnetic field that wraps around the direction of the current. A moving electric charge or electric current creates a magnetic field perpendicular to the charge's movement. Many neurons in the computer brain's circuitry are bent, and this creates a magnetic field inside the neuron. If the magnetic field inside of the neuron experiences change, it creates an electric current in the neuron and causes the electric charges to move. An electric charge is a physical property of normal matter that causes the computer brain to experience a force when placed in an electromagnetic field.

The inhabiter's focal point of dark energy and peripheral surface of enabled imprint strings communicate with each other and yield energy as moving magnetic charges. Oscillating enabled imprint strings are magnetic monopoles, meaning they are isolated magnets with only one magnetic pole, and they create an electric field by means of a moving magnetic charge. An enabled non-ego-clinging imprint string is a positive magnetic monopole, and an enabled ego-clinging imprint string is a negative magnetic monopole.

As dancing strings of energy, enabled ego-clinging and non-ego-clinging imprint strings are in actuality impossible to see, but their appearance can be conceptualized by means of the psyche. The inhabiter's focal point of dark energy possesses background energy, and this is received by the enabled imprint strings as magnetic energy that causes the strings to oscillate and thereby function. Enabled imprint strings that oscillate and are charged with magnetic energy possess a magnetic flowing current. A moving magnetic charge creates an electric field in a circle around a straightened, oscillating, enabled imprint string. Magnetic charges move through an enabled imprint string when an electric field gets changed.

Ego-clinging and non-ego-clinging imprint strings are connected to the focal point of dark energy and located on its peripheral surface, but imprint strings oscillate differently. When imagining what enabled imprint strings look like, there are times when their appearance is similar, but when non-ego-clinging imprint strings tap in, the appearance of enabled imprint strings are dissimilar. Both types of the inhabiter's imprint strings were meant to fulfill a purpose and do so by means of the guidance they receive.

When enabled non-ego-clinging imprint strings tap in, they fluctuate in a way so that they bend and thereby receive guidance from the inhabiter's focal point of dark energy. While bent, the magnetic current flowing through an enabled non-ego-clinging imprint string produces an electric field inside the imprint string. Enabled ego-clinging imprint strings receive guidance from the computer brain's meta-algorithm, which allows the inhabiter to procure an immersive experience while interacting with normal matter. Ego-clinging imprint strings have a magnetic current flowing through them, but because they do not bend, an electric field is not present inside the imprint string; rather, the electric field only forms a circle around the enabled ego-clinging imprint string.

With the universal law of cause and effect in force, what the inhabiter does with its enabled imprint strings affects the brain's circuitry—and the converse is also true. The fluctuating electric fields are always associated with magnetic fields that change over time. Correspondingly varying magnetic fields are associated with specific changes over time in the electric fields.[239] The normal matter computer brain and the inhabiter both produce electromagnetic radiation, which is waves of energy that travel at the speed of light. Electromagnetic radiation consists of electromagnetic waves that are synchronized oscillation of electric and magnetic fields. The frequency of electromagnetic radiation refers to the number of times the waves peak and trough in a given second, and the electromagnetic wave could be classified either by its frequency of oscillation or its wavelength. Photons are packets of energy of an electromagnetic wave.

In an electromagnetic wave, the changing magnetic field is always associated with a changing electric field and vice versa.[240] Magnetic fields may be viewed as relativistic distortions of electric fields, and together these fields form propagating electromagnetic waves.

The computer brain's firing neurons produce electrical energy that the meta-algorithm transmits throughout the computer brain and network as signals of electric currents of varying intensities.

A normal matter computer brain configured to generate data for a human psyche has thousands of neurons that fire simultaneously at the same frequency and generate electromagnetic waves with a rate of ten to one hundred cycles per second.[241] The normal matter computer brain produces electromagnetic radiation and this radiation is the means by which the computer brain's information is transferred as continuous signals to the inhabiter's enabled imprint strings.

Although the computer brain's claustra receives input from the cortical areas of the computer brain via five different tracts and the meta-algorithm created from claustral programs detects neural synchrony, undoubtedly, the computer brain could not function as a unit without the spinal cord and brain stem. The spinal cord connects nearly all parts of the computer brain's lower regions with its upper regions. The brain stem has programs that compute for the brain's life-regulation devices and is located between the cerebral hemispheres and the spinal cord. The meta-algorithm is the computer brain's self-contained transmitter and receiver of information, but the brain stem and spinal cord function as the normal matter computer brain's "antenna."

The spinal cord, part of this antenna, is a cylindrical bundle of nerve fibers with electric charges flowing through it; these currents generate magnetic fields perpendicular to the charges' movement. Spinal nerves are paired on a common axis (the spinal cord), and each nerve has an electric current flowing through it producing a magnetic field around the nerve. Each pair of nerves connects the spinal cord with a specific region in the normal matter computer brain. Due to the paired nerves' proximity, a magnetic field of one nerve passes through the other, and a change in the electric current though one spinal nerve induces a voltage (electromotive force) across the other.[242] Accelerating electric charges traveling throughout the spinal cord produced by neurons firing at different frequencies create time-varying electric and magnetic fields in the space around them. In the region of the brain stem, there are vacillating electric and magnetic fields, and the computer brain's information is transferred as signals by electric fields. When signals from the computer brain are received by enabled imprint strings, it changes the way they oscillate.[243]

The inhabiter's created cognitive states are contained within the electromagnetic radiation[244] of the psyches that travels like an invisible wind[245] across the surface of the inhabiter's focal point of dark energy. This electromagnetic wind is light and powerful[246] and is the product of the interplay between the inhabiter's enabled imprint strings and the normal matter computer brain's electrical patterns.

The electromagnetic radiation created by the inhabiter's imprint strings carries two neurointeractive psyches about the inhabiter's life, but from the inhabiter's viewpoint, its life story has the appearance of one version. There is the version created by enabled ego-clinging imprint strings, entitled "psyche self-centered," and the version created by enabled non-ego-clinging imprint strings, which displays the updated and accurate alternative is entitled "psyche altruistic." Both versions are displayed in the electromagnetic radiation, but how much of a particular version the inhabiter perceives depends on the power of the forces within the electromagnetic radiation.

When imprint strings are disabled, a single photon of light is produced,[247] and both types of imprint strings, when enabled, have opposite charges. As photons are produced by the disabling of imprint strings, they will affect the forces within the electromagnetic radiation and correspondingly the inhabiter's perception. The ability of the inhabiter to switch from the immersive task of watching and responding to psyche self-centered to that of purely perceiving psyche altruistic is based on the forces in the wind. The version driven by the most enabled imprint strings carries the greatest reciprocal charge, which correlates with the ability it has to influence the inhabiter's perceptions and beliefs.

Enabled imprint strings function as the transmitters of magnetic energy as alternating signals of magnetic charges and the receivers of signals from the computer brain's meta-algorithm in the form of electric currents of varying intensities. Oscillating, enabled imprint strings produce electromagnetic radiation and use it to transmit information to the computer brain, specifically to the region of the interclaustral pathway where the claustra share their data and the meta-algorithm selects and modifies timing of synchrony so agents are fashioned and linked. The electromagnetic radiation has within it electric and magnetic fields, and there are two sources of electromagnetic radiation that travel on the surface of the inhabiter's focal point of dark energy: enabled ego-clinging imprint strings and enabled non-ego-clinging imprint strings create the electromagnetic radiation of the psyche self-centered and the psyche altruistic.

Though the computer brain regularly sends a plethora of information as signals, enabled imprint strings have the ability to select information sent by normal matter, discriminating which data is needed to create a sense impression or mental event. Both psyche self-centered and psyche altruistic are comprised of sense impressions and mental events, and the psyches are created via four steps:

- step one: signals received
- step two: signals transformed
- step three: primordial awareness
- step four: feeling tones and response

However, in order for enabled imprint strings to completely devise the inhabiter's created cognitive states, two rotations of the four step sequence are required, one following directly after the other: during the first rotation, the computer brain's data is selected and transformed by enabled imprint strings to create the sense impressions of "gross" consciousness present within psyche self-centered and psyche altruistic.

While interacting and transforming the signals received by the normal matter computer brain, the vibrations of the inhabiter's infinitely small strings of energy (enabled imprint strings) will create the illusory phenomenal world that the inhabiter will experience through the I character's senses. Sense impressions are physical sensations that do not include images; rather, they are sensations within the psyche, and by means of created cognition, the inhabiter will quantify its sensibility of the illusory phenomenal world using descriptive names such as the following:

- earth—solidity
- air—movement
- water—moisture
- fire—temperature

As an alternative, the inhabiter may use scientific nomenclature such as the fundamental forces of nature:

- gravity
- electromagnetism
- strong force
- weak force

Whichever the case might be, the sense impressions contain physiologic information about the computer brain that embodies the inhabiter and comprises gross consciousness.

As information processing continues during the second rotation of the four steps, the computer brain's data is transformed further by enabled imprint strings to create mental events. Mental events are the thoughts, emotions, and memories present within the psyches and are termed "subtle" consciousness.

Step One: Signals Received

Enabled imprint strings selectively choose data based on whether they are creating a sense impression or mental event according to the four steps. During the first rotation, to create sense impressions of gross consciousness, enabled imprint strings will select the data that pertains specifically to the normal matter computer brain. Amid the second rotation, while constructing mental events present within subtle consciousness, enabled imprint strings select the information received by normal matter from the network of interconnected computer brains spread throughout the universe.

Although imprint strings have never been directly seen or measured, one significant discrepancy between enabled ego-clinging imprint strings and non-ego-clinging imprint strings is that they oscillate differently, which causes them to produce electromagnetic waves of energy as electromagnetic radiation that have divergent frequencies. The psyches created by enabled imprint strings consist of electromagnetic waves. Since the inhabiter has two types of imprint strings enabled (ego-clinging and non-ego-clinging), there will be two psyches displayed on the inhabiter's focal point of dark energy: psyche self-centered and psyche altruistic correspondingly.

When signals are received from the normal matter computer brain, the electromagnetic waves of both psyches are affected as the electric fields gets changed. Imprint strings as moving magnets oscillate differently, which diversifies content within psyche self-centered and psyche altruistic.

Step Two: Signals Transformed

Oscillating enabled imprint strings are arranged as pairs of ego-clinging and non-ego-clinging imprint strings on the inhabiter's focal point of dark energy. Each enabled imprint string has flowing through it a moving magnetic charge that creates an electric field perpendicular to the magnetic charge movement. Imprint strings are located on the peripheral surface of the inhabiter's focal point of dark energy and are close in proximity. An electric field of one enabled imprint string passes through the other, and there are time-varying magnetic and electric fields in the area around the strings.

Upon receiving and selecting signals generated by the computer brain, enabled ego-clinging imprint strings change their oscillations in a way that is directed by the claustra's meta-algorithm. When non-ego-clinging imprint strings receive and select the computer brain's signals, initially they will oscillate in a way that is similar to ego-clinging imprint strings and therefore according to how the brain instructs. Nonetheless, what enabled non-ego-clinging imprint strings do next alters the way they oscillate because they bend. As non-ego-clinging imprint strings bend, the electric field around the imprint strings is modified; instead of an electric field in a circle around the imprint strings, the electric field relocates inside the imprint strings, and when fully bent, both ends of the enabled imprint strings are connected to the inhabiter's focal point of dark energy. Thus, the electric field inside the enabled non-ego-clinging imprint strings carries information about the meta-algorithm's processing as synchrony fits, and this is made known to the dark energy focal point within the inhabiter's configuration in an instant. Not directly connected to the meta-algorithm, the inhabiter's focal point of dark energy relies on enabled non-ego-clinging imprint strings to share information they received from the normal matter computer brain.

When enabled non-ego-clinging imprint strings are tapped in, both of their ends are connected to the focal point of dark energy, and there will be a magnetic current flowing through both ends of the imprint strings. Additionally, the enabled non-ego-clinging imprint strings will have an electric field running through it in opposite directions. The electric fields inside each bent and tapped-in, enabled non-ego-clinging imprint string are equal in magnitude but are traveling in opposite directions; therefore, the fields cancel each other out, since adding two fields that are equal and opposite results in zero. This has a powerful effect on the enabled non-ego-clinging imprint strings by decreasing their electrical reactivity,[248] which allows dark energy to guide them through changes in the modulation of their oscillations. The moment the enabled imprint strings tapped into dark energy, the computer brain's data was analyzed, and the imprint strings then disconnect from the focal point at one end. When no longer tapped in but still connected, the enabled non-ego-clinging imprint strings will have received guidance from the focal point of dark energy and will oscillate in a way that differs from that of ego-clinging imprint strings.

Though "gross" and "subtle" consciousness are within both psyches, enabled imprint strings will create sense impressions before mental events. This means sense impressions created will have completed the four-step sequence before the electromagnetic waves used to conceive them become the building blocks by which enabled imprint strings create mental events.

After enabled imprints strings create sense impressions at the peripheral surface of the focal point of dark energy, the electromagnetic radiation of both psyches travels from the focal point's peripheral surface, where the enabled imprint strings are located, to the interior surface of the focal point, and the inhabiter becomes aware of the sense impressions. The movement of electromagnetic radiation on the surface of the focal point of dark energy is due to the kinetic energy produced by the oscillations of enabled imprint strings. When enabled imprint strings oscillate, kinetic energy is produced that moves electromagnetic radiation of both psyches from place to place on the surface of the focal of dark energy. The slower the enabled imprint strings oscillate, the momentum by which the electromagnetic radiation moves will be decreased, and there will be less kinetic energy present on the surface of the focal point of dark energy. The converse is also true: the faster enabled imprint strings oscillate, the more electromagnetic radiation moves with an increased speed, and there will be more kinetic energy present on the surface of the focal point of dark energy.

Step Three: Primordial Awareness

In step three, whether occurring during the first or second rotation of the four-step sequence, the inhabiter uses its primordial consciousness derived from its focal point of dark energy and becomes aware of the electromagnetic waves of both psyche self-centered and psyche altruistic.

The four-step sequence used by enabled imprint strings to create sense impressions and mental events of gross and subtle consciousness cannot manifest without the computer brain's information and the data received by the brain from the network of connected normal matter computer brains. Be that as it may, there is another type of consciousness present within the inhabiter's configuration that is not created but rather simply exists, and it is called "extremely subtle" consciousness.

Extremely subtle consciousness is not created by enabled imprint strings via the four-step sequence; ergo, it is neither gross nor subtle consciousness. This consciousness is instead the pure awareness of the inhabiter's focal point of dark energy that is embodied within the normal matter computer brain. Extremely subtle consciousness is awareness but is void of all physical sensations, active thoughts, emotions, and memory and is imbued with three attributes: bliss, luminosity, and nonconceptuality.[249] In this way, extremely subtle consciousness may be characterized as the ground state of the psyche,[250] and generally speaking, it is indiscernible by the inhabiter while the psyche is active, for it normally manifests in dreamless sleep and during the last moment before brain death.[251] The inhabiter therefore cannot experience extremely subtle consciousness" by investigating fleeting sense impressions or mental events. Thereupon, the inhabiter is limited to simply acknowledging the focal point of dark energy's presence within its configuration without a subject/object reference. The Inhabiter's focal point of dark energy is the source that powers enabled imprint strings allowing them to function, and illuminates the inhabiter's created cognition within both psyches in a way that allows the inhabiter to interact meaningfully with the normal matter computer brain.

Electromagnetic waves within the electromagnetic radiation of the psyche self-centered and psyche altruistic contain information of sense impressions and mental events, but the inhabiter's focal point of dark energy does not see it that way. The primordial consciousness of dark energy is pure awareness that is not fabricated and is devoid of the appearances of created conscious states. Dark energy as a focal point analyzes information with nondualistic perception, evaluating only the essential information as physical attributes[252] incorporated within the electromagnetic radiation of both psyches. This includes charges,[253] magnetic and electric forces of electromagnetic waves, and the mass of normal matter as the computer brain that embodies the inhabiter. The inhabiter's focal point of dark energy analyzes information because the ability it has to guide enabled non-ego-clinging imprint strings and use them to alter the claustra's activity is dependent on the electromagnetic radiation's power and frequency.[254]

Step Four: Feeling Tones and Response

In step four, the inhabiter is using the primordial consciousness of its focal point of dark energy to illuminate the electromagnetic waves created by its enabled imprint strings, and thereby the inhabiter will think with created cognitive states. However, during the first rotation of the four-step sequence, the inhabiter's conscious states are not yet fully developed, which means the inhabiter will respond to the sense impressions (physical sensations) of gross consciousness by predominantly reacting as a response versus cognizing its choices. The inhabiter aware of sense impressions will associate with them one of three different types of *feelings tones*. A feeling tone is not an emotion such as fear, joy, or anger but is rather the immediate and spontaneous affective experience of the inhabiter's awareness of a sense impression.[255]

Feeling tones are one of the following:

- pleasant
- unpleasant
- neutral

The feeling tones the inhabiter associates with sense impressions contained within the electromagnetic waves of both psyches in step four, during the first rotation, will be associated with more complex events in step four, during the second rotation, as the inhabiter's cognitive states become fully developed and will influence the inhabiter's choices and responses to both psyches through the primary emotions (happiness, sadness, fear, anger, and surprise) and the four secondary emotions (embarrassment, jealousy, guilt, and pride).

The inhabiter's response to sense impressions with associated feeling tones is automatic and often unnoticed by the inhabiter. But when the inhabiter responds to a sense impression with an associated feeling tone, a reverberation is produced that signifies the inhabiter's response. The repercussion begins on the interior surface of the focal point of dark energy and travels to the peripheral surface. When the force of the reverberation reaches the peripheral surface of the dark energy focal point, this signals enabled imprint strings to use the sense impressions with associated feeling tones to make mental events. The enabled imprint strings oscillate in a way so that kinetic energy is produced on the surface of the focal point that moves the electromagnetic radiation of both psyches with sense impressions and associated feeling tones from the interior surface of the focal point of dark energy to the peripheral surface, where the imprint strings are located.

Enabled imprint strings are able to oscillate so as to propagate new information among one another to create mental events and yet retain the information they previously created as electromagnetic waves representing sense impressions. In this way, sense impressions are the building blocks for the creation of mental events and contain specific physiologic data about the computer brain that will be used by enabled imprint strings to create the proto-I character as the foundation used to construct the core or autobiographical-I character.

Second Rotation of the Four-Step Sequence

Step One: Signals Received; Step Two: Signals Transformed
The enabled imprint strings repeat the initial two steps of the four-step sequence to make mental events, similar to the way they created sense impressions. They do so, however, with a slight discrepancy. The disparity between the actions enabled imprint strings take when creating sense impressions versus mental events is this: rather than starting the sequence with the computer brain's signals, enabled imprint strings use

the information contained within the electromagnetic waves of sense impressions with associated feeling tones as a building block. Adopting sense impressions with associated feeling tones as a building block, enabled imprint strings then select the signals coming from the computer brain that contain information received from the network of normal matter computer brains. Consequently, the data that enabled imprint strings will use to create a mental event and subtle consciousness involves more than what was used to create sense impressions and gross consciousness. This means mental events and subtle consciousness contains more additional information than what primarily pertains to the computer brain that embodies the inhabiter. Enabled imprints strings, having expanded the information used to create sense impressions, will correlate data to generate mental events, which includes the creation of the I character.

Enabled imprint strings begin constructing the I character using electromagnetic waves representing sense impressions that contain information about the brain's physiologic processes, and the proto-I will be the foundation for either the core-I or the autobiographical-I character. The I character is the computer brain's optimally intelligent agent that acts as a centralized control of the claustra's processing, and the proto-I character is vitally important for the computer brain's survival. However, the core-I character and the autobiographical-I character are created by the computer brain for different reasons. The computer brain creates the core-I character when not processing historical computations, but in order for the claustra to continue to process information efficiently, they need access to the history of their inputs and actions and therefore will generate the autobiographical-I[256] character. The enabled imprint strings create their respective psyches with the information they receive from the computer brain, and two rotations of the four-step sequence represent the inhabiter's cognitive cycle. Each of the inhabiter's cognitive cycles will contain either a core-I character or autobiographical-I character, not both, and enabled ego-clinging and non-ego-clinging imprint strings, having received the same signals from the computer brain, will create the same character. However, whichever version of the I character enabled imprint strings create within psyche self-centered or psyche altruistic will not be perceived by the inhabiter until step four of the second rotation of the four-step sequence.

Once the I character is created, by means of a complex series of oscillations, more electromagnetic waves and patterns will be produced by enabled imprint strings that generate the phenomena of the illusory world within the psyches that surround the I character.

At the completion of step two, the electromagnetic radiation of both psyches will again travel from the peripheral surface of the focal point of dark energy to the interior surface, and using its primordial consciousness, the inhabiter will become aware of the electromagnetic radiation of both psyches.

Step Three: Primordial Awareness

The gift of the inhabiter's configuration is that enabled imprint strings and the focal point of dark energy work collaboratively to create the inhabiter's reality while embodied within a normal matter computer brain. Enabled imprint strings function as data transformers, and the focal point of dark energy is the inhabiter's intelligence that is not devised but rather simply is.

During step three, the primordial intelligence of the inhabiter's focal point is alertness that analyzes the electromagnetic waveforms within both psyche self-centered and psyche altruistic simultaneously. Its interpretation is coherent, allowing dark energy to guide enabled non-ego-clinging imprint strings in adjustments to their oscillations.

Step Four: Feeling Tones and Response

Unlike step four in the first rotation, the inhabiter's response to a mental event with an associated feeling tone is often not ad lib, because the inhabiter's cognitive states are fully developed. The inhabiter's habitual reaction to a mental event with an associated feeling tone is to pursue those that are pleasant and to avoid those that are unpleasant; these are known as attachment and aversion respectively.[257] It is the inhabiter's habitual reactions of attachment or aversion to the feeling tone associated with a mental event (thought, memory, emotion) that produce the inhabiter's desire or repulsion[258] and is a nuance of cause and effect. The inhabiter's attachment and aversion arise in its reaction to the feeling tone itself, rather than to a thought, memory, or emotion (mental event) that compose subtle consciousness present within the psyche.[259]

Although electromagnetic radiation is within the infrared spectrum, the inhabiter is able to perceive information contained within both psyches because the inhabiter is using its dark energy primordial consciousness to illuminate electromagnetic waves. The inhabiter's dark energy focal point grants the inhabiter the gift of perception, enlivens the inhabiter's created conscious states to make them meaningful, and allows the inhabiter to witness what normally would be invisible.

What is important to consider is that enabled ego-clinging and non-ego-clinging imprint strings create dozens of sense impressions and mental events as a rapid series in a given second, so the inhabiter's cognitive states are created via the four steps at an extremely rapid rate that is indiscernible to the inhabiter. The inhabiter will experience a continuous stream of created consciousness that begins with a feeling tone, is followed by a sense impression and then a mental event. The Inhabiter will experience mental events as a rapid succession of three-dimensional images or holograms, the content of a psyche produces the illusion of movement in the display presented to the inhabiter on the interior surface of its focal point of dark energy. By virtue of the I character's body, the inhabiter will experience sights, sounds, smells, tastes, and touch, and the world within both psyches becomes, from the inhabiter's viewpoint, "real."

Even though the inhabiter is using its primordial consciousness to illuminate the significance of the computer brain's data, during step four, the inhabiter thinks with its created cognitive states. Psyche self-centered and psyche altruistic are within electromagnetic radiation that has divergent frequencies, and each psyche contains cognitive states that have noncoherent viewpoints. This means the inhabiter will experience interference within its created cognition as it relates to the inhabiter's perception and beliefs.

The inhabiter selects one of the two psyches to experience its momentary reality, and in doing so, the inhabiter thereby is responding to the psyche in a way that disables and enables its imprint strings. The inhabiter's awareness of objects within both psyches will be accompanied by associated feeling tones. If the inhabiter's response is motivated by either attachment or aversion, the inhabiter will choose the psyche self-centered. Whereas if the inhabiter's response is motivated by a disinterest in using feelings to guide its choices, and instead, the inhabiter's goal is to transform itself for benefit, the inhabiter will select the psyche altruistic. Whichever the case might be, the electromagnetic radiation of both psyches dissolves into the focal point of dark energy, and this causes a reverberation with a force that correlates directly with the inhabiter's motivation.

Subsequently, the reverberation moves across the surface of inhabiter's focal point of dark energy until it reaches the enabled imprint strings and will determine the type of imprint strings and how many of them will be disabled.[260] As the strings are arranged on the inhabiter's focal point of dark energy as ego-clinging and non-ego-clinging pairs, a change in a magnetic charge through one type of enabled imprint string causes a magnetomotive force across the other that disables or enables the other type of imprint string. Specifically, an ego-clinging imprint string disables a non-ego-clinging imprint string, and the opposite mechanism is at work when non-ego-clinging imprint strings are enabled. The enabled imprint strings that are left will transmit alternating signals of magnetic charges, which will be picked up by the normal matter computer brain's antenna of the brain stem and spinal cord and will be received by the meta-algorithm in the interclaustral pathway.

In this way, each enabled imprint string gives the inhabiter programming access[261] to either follow or alter how the meta-algorithm modulates and links agents to the I character as it designs the computer brain's virtual world around this anomaly. To the degree enabled ego-clinging imprint strings lock the inhabiter into following the directives of the meta-algorithm, the inhabiter's perception is primarily focused on the psyche self-centered, and its ability to perceive reality accurately and operate as a skilled participant in its existence embodied is minimal. It is only the enabled non-ego-clinging imprint strings that create the back doors[262] of paradoxical programming that the meta-algorithm would not perform on its own. Each positive charge of the inhabiter's enabled non-ego-clinging imprint strings holds a code expressed as vibrational patterns, allowing them to move and tap into the inhabiter's focal point of dark energy that will guide these imprint strings in a way that benefits the inhabiter and the normal matter computer brain.

The focal point of dark energy, having acquired the electromagnetic radiation of both psyches, transforms it to a discrete quantity of energy as a signal to the universe's dark energy substrate layer that is proportional in magnitude to the frequency of the radiation it represents.

The universe's dark energy substrate layer acts as a two-dimensional grid, and within that grid are an unknown number of attached focal points of dark matter. Albeit the exact number of focal points is unknown, the dark energy substrate layer does not perceive them as being separate from itself, because they are not. Focal points of dark energy within every inhabiter's configuration are the same dark energy from the universe's substrate layer; with a surface of imprint strings, dark energy is simply transformed to its interactive form, dark matter.

Each inhabiter embodied by a normal matter computer brain is still attached to the universe's dark energy substrate layer. Because there are many differently configured normal matter computer brains that produce disparate data spread throughout the universe, many inhabiters will be inhabiting divergently configured computer brains. While embodied, the inhabiter's enabled imprint strings will use the data generated by the computer brain to create their respective psyches; this means inhabiters will experience life with similar and dissimilar species of living beings or I characters, and their stories of conditioned existence and experiences will be unique. The psyches are riddled with symbolic representations: the I character represents the inhabiter's direct interaction with normal matter and with the character inhabiters derive a sense of self. The sense of self inhabiters experience through the I character is utilized by the universe's dark energy substrate layer, which elaborates the concept by facilitating the communication between inhabiters and is symbolically represented within psyche self-centered and psyche altruistic as other living beings. When inhabiters are corresponding, their psyches are said to be entangled. Dark energy as the universe's substrate layer is not interacting with normal matter; rather, it simply explores what this interaction is like in many different ways. This is its means to gather pertinent information, identify paths of alternative actions, and discern the consequences of these actions in order to answer this question: There are myriad ways to occupy the universe and coexist. Both dark energy and normal energy can occupy the universe without interacting or can interact as dark matter inhabiters and normal matter computer brains. Which of the two options results in universal bliss?

Dark energy as the universe's substrate layer will relay signals to many inhabiters spread throughout the universe simultaneously no matter how far apart they are. The concepts of time and space are limitations created within psyche self-centered and psyche altruistic, which have no effect on the universe's dark energy substrate layer that is facilitating communication between focal points of itself. The signals sent by the universe's dark energy substrate layer contain information about inhabiter's I character, and each inhabiter communicating will be receiving information at the same time, which creates the illusion in the psyche to the inhabiter observing it that its I character is interacting with other living beings. Because I characters and other living beings have no actual reality, their characteristics do not belong to them intrinsically; rather, the inhabiter's perceptions will be based on its cognitive states created by enabled imprint strings using the data generated by the computer brain that embodies the inhabiter. If the inhabiter is embodied by a normal matter computer brain, with minimal capacities to produce data for cognitive states though a signal is received by a focal point from the universe's dark energy substrate layer, a living being will not be conceptualized by the inhabiter.

As quantum information courses through the labyrinth of the universe's dark energy substrate layer as signals, where the data travels produces a path that heats up a region within the substrate layer to an excited state, while the rest of the universe's dark energy substrate layer is background energy in a naturally grounded tranquil state. Notwithstanding, signals are received by focal points of dark energy during the second rotation of the four-step sequence and during step two while enabled imprint strings are creating mental events. Enabled imprint strings, having received the information, will oscillate in a very unique way that differs from the way they oscillate when transforming the computer brain's data to create their respective psyches. However, enabled imprint strings receive signals during the period when they are diligently creating their mental events within their respective psyches with the computer brain's data. Nevertheless, the information received by the universe's dark energy substrate layer concerns only information about different inhabiter's I characters, and the enabled imprint strings will know the information has been received from the universe's substrate layer, not by alterations in scenery but rather through a local type of electromagnetic field.[263] This is a local type of electromagnetic field produced when an inhabiter's focal point of dark energy is receiving signals from the universe's dark energy substrate layer.

Enabled imprint strings create doors to mysterious programming not initiated by the normal matter computer brain that embodies the inhabiter but rather by the universe's dark energy substrate layer. The information received by the focal point of dark energy from the universe's substrate layer will be incorporated contextually into both psyche self-centered and psyche altruistic in a way that will make sense to the inhabiter because it complements mental events. However, the inhabiter will not perceive the I character, other living beings, or phenomena until its cognitive states are fully constructed in the second rotation of the four-step sequence and during step four.

Dark energy orchestrates entanglement with omnipotence and nonconceptual awareness that knows itself and which inhabiters are communicating and thereby constructs its own kind of map of the nonseparable universe on a cosmic scale. Dark energy's ability to scatter quantum information as signals to different inhabiter's focal points of dark energy may seem like a *spooky* action for dark energy to take, yet dark energy, whether part of an inhabiter's configuration or in the form of the universe's substrate layer, depends on all parts of itself to be complete and undivided. It is a mystery as to when the universe's dark energy substrate layer will send a signal to a focal point of dark energy. If a focal point of dark energy within an inhabiter's configuration does not receive a signal from the universe's dark energy substrate layer, the inhabiter will only perceive the psyches its enabled imprint strings create. This means the inhabiter may perceive an I character where there will be no other living beings present within the psyche. Instead, the information that enabled imprint strings access and interpret will be that from the computer brain and what the brain receives from the network of connected normal matter computer brains, and enabled imprint strings will create the psyche that includes an I character surrounded by inanimate things within an ambient environment.

At the conclusion of the inhabiter's cognitive cycle, enabled imprint strings will send signals to the computer brain's interclaustral pathway, and therefore novel information will be incorporated into interconnected networks that compute and store information. The claustra access this with their widely distributed anatomic projections that extend to almost all regions of the normal matter computer brain's cortex and many of its subcortical structures. The meta-algorithm adds on a lot of information to what is going on in the signals that come from the inhabiter's enabled imprint strings, and the computer brain plots signals generated by its circuitry as neural maps. Therefore, the inhabiter's communication with another inhabiter becomes a part of the computer brain's history, and even when an inhabiter is not actively communicating with another inhabiter, the computer brain has the interaction stored in its memory circuits and will relay this data as signals back to the inhabiter's enabled imprint strings during times when the meta-algorithm is sequencing data for the autobiographical-I character. The inhabiter will experience this within the psyche as the memory of the I character interacting with a particular living being or many living beings and amidst the inhabiter's anticipated future contained within its cognitive states.

Dark energy designed ego-clinging imprint strings for a purpose. The enabled ego-clinging imprint strings that create the inhabiter's immersive experience with the cognitive states of the psyche self-centered are necessary and part of the cycle of life while embodied within a normal matter computer brain. How long will this cycle last? Only the inhabiter can answer this question, because it is dependent on the inhabiter's responses to the psyche and its choices, as the inhabiter has the ability to alter the computer brain's programming using its primordial dark energy consciousness. This means that using its intelligence that is not created, the inhabiter will need to train itself to be a skillful observer, and it can learn to do this by choosing the psyche altruistic. How much time the inhabiter has to train itself is unknown, but the clock is ticking. From the moment the inhabiter began inhabiting the computer brain, it began its unpredictable but inevitable computation for impermanence.

The omnipotent universe's dark energy substrate has a boundless scope of awareness of each inhabiter and the population of enabled and disabled imprint strings. However, dark energy's awareness transcends all distinctions of subject and object, mind and matter, and words and concepts.[264] Assuredly, uncertainty underlies entanglement, and the complexity of the unfolding is within the realm of the universe's dark energy substrate layer; after all, it is dark energy that facilitates the communication with the parts of itself that are embodied as inhabiters. Entangled inhabiters participate with one another through the psyche and will exchange information in a variety of different ways; clearly, it is through interactions with one another that inhabiters will enrich their realities while embodied within a normal matter computer brain.

This means the universe's dark energy substrate layer does not make a list of the inhabiters that are naughty or nice, nor does dark energy punish an inhabiter or determine the inhabiter's fate. Undoubtedly, it is the inhabiter that ultimately punishes itself and determines its fate, and the universe's dark energy substrate layer, as the orchestrator of entanglement, will deliver to the focal point within the inhabiter's configuration signals that contain information that allows the inhabiter to self-actualize its choices.

Key Point

- The inhabiter trains itself in accurate perception when it continually reminds itself that an independently existing I character has no actual reality and is only a mental construct within the psyche.

Enabled ego-clinging imprint strings are negative magnetic monopoles that have a magnetic current flowing through them and an electric field that forms a circle around the enabled imprint string. This image depicts the electric and magnetic fields generated in electromagnetic waves.

In the image, the large gray line represents the magnetic current flowing through an ego-clinging imprint string. The white circles represent the electric fields, and the gray circles represent the magnetic fields.

Electric and magnetic fields in electromagnetic waves
created by enabled ego-clinging imprint strings.

Enabled non-ego-clinging imprint strings are positive magnetic monopoles that will initially oscillate similar to enabled ego-clinging imprint strings until they tap in by bending and thereby receive guidance from the inhabiter's focal point of dark energy. While bent, the magnetic current flowing through an enabled non-ego-clinging imprint string produces an electric field inside the imprint string and creates electromagnetic waves.

In the image, the gray line represents the magnetic current flowing through a bent, enabled non-ego-clinging imprint string, and the white lines represent the electric field produced.

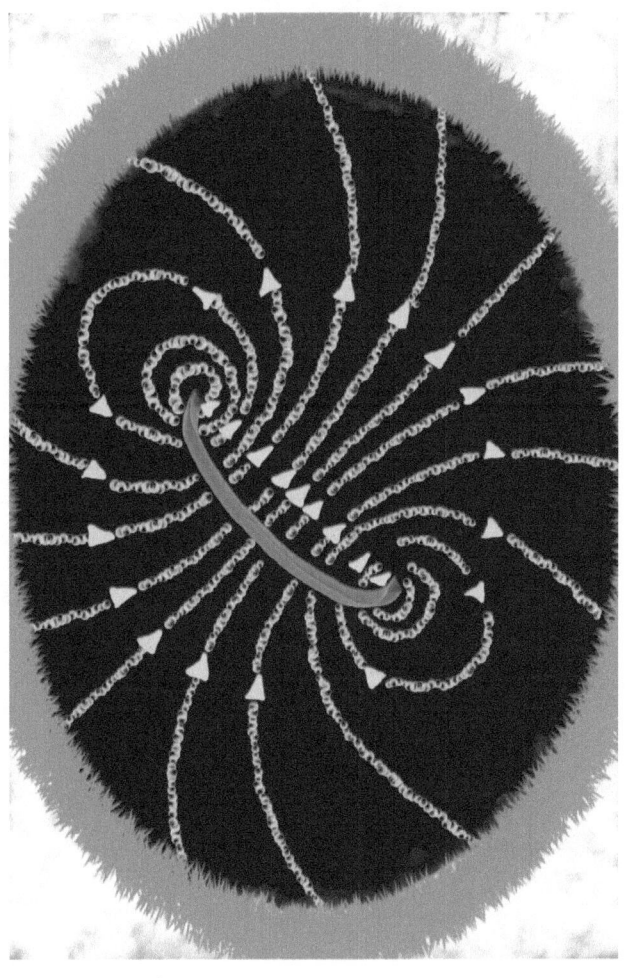

Electric and magnetic fields in electromagnetic waves created
by enabled non-ego-clinging imprint strings.

Electromagnetic radiation consists of electromagnetic waves, and the inhabiter's created cognitive states are contained within the electromagnetic radiation of the psyche self-centered and the psyche altruistic.

The image displays electromagnetic waves.

Electromagnetic waves.

Reminders before Reading Chapter 10

Electromagnetism and Related Topics

- electromagnetism: A fundamental force.
- electromagnetic radiation: A kind of radiation in which electric and magnetic fields vary at the same time.
- electromagnetic waves: Synchronized oscillation of electric and magnetic fields.

Four Steps to the Creation of the Psyches

- psyches: Comprised of sense impressions and mental events and are created via four steps.
- four steps: Steps involved in the creation of the inhabiter's cognitive states—signals received (step one), signals transformed (step two), primordial awareness (step three), feeling tones and response (step four).
- sense impressions: Physical sensations.
- gross consciousness: Consists of sense impressions and contains physiologic information about the normal matter computer brain that embodies the inhabiter.
- mental events: Thoughts, emotions, and memories present within the psyches, displayed as a rapid succession of three-dimensional images or holograms.
- subtle consciousness: Consists of mental events and contains information received from the network of normal matter computer brains.
- feeling tone: An immediate and spontaneous affective experience of pleasant, unpleasant, or neutral, and it is not an emotion such as fear, joy, or anger.
- extremely subtle consciousness: Awareness, but is void of all physical sensations, active thoughts, emotions, and memory.

Imprint Strings

- enabled ego-clinging imprint strings: Have a magnetic current flowing through them and an electric field that forms a circle around the enabled imprint string.
- enabled imprint strings: Magnetic monopoles and isolated magnets with only one magnetic pole. They create an electric field by means of a moving magnetic charge.
- enabled non-ego-clinging imprint strings: Initially oscillate similar to enabled ego-clinging imprint strings until they tap in by bending and thereby receive guidance from the inhabiter's focal point of dark energy. While bent, the magnetic current flowing through an enabled non-ego-clinging imprint string produces an electric field inside the imprint string.

Miscellaneous Topics

- entanglement: A phenomenon orchestrated by dark energy that manifests when information received by a focal point of dark energy from the universe's substrate layer is incorporated contextually into both psyche self-centered and psyche altruistic in a way that will make sense to the inhabiter because it complements mental events.
- focal point of dark energy: Possesses background energy that is received by the enabled imprint strings as magnetic energy, which causes the imprint strings to oscillate and thereby function.
- neuron: Electrical cell that produces an electric current that induces a magnetic field that wraps around the direction of the current.
- universe's dark energy substrate layer: Acts as a two-dimensional grid, and within that grid are an unknown number of attached focal points of dark matter. As the substrate layer, dark energy relays signals of quantum information about the inhabiter's I character to focal points of dark energy.

CHAPTER 10

The Primordial Beam and the Dream or Nightmare

Dark energy, in its noninteractive form, fills the universe and is space that is devoid of created conscious states, but its boundless awareness is enduring. Normal energy, when not interacting with dark energy in the universe, does not integrate into a normal matter computer brain that creates electrical patterns and is undecidedly free of impermanent states of existence. Fully aware of its presence and that of normal energy in the universe, dark energy understood the ramifications before beginning its interaction with normal energy. In order to overcome any barriers that might hinder its efforts to live harmoniously with normal energy, dark energy ascertained that it must engage normal energy to attain vital information. Nonetheless, there is much more dark energy in the universe than normal energy; thus, if dark energy engaged normal energy during one interaction, allegorically, normal energy would be overpowered and swallowed up by the experience. Therefore, the universe's dark energy substrate layer designed the interaction with normal energy in a way that accommodated the inequivalence of power. Dark energy morphed imprint strings and transformed itself to dark matter, thereby constructing the variables in an equation of interaction that set into motion the transformation of normal energy into a normal matter computer brain. Tolerance and uniformity in the distribution of power are principles followed by the universe's dark energy substrate layer that simply guides itself according to its compassionate nature. Hence, dark energy is not a dominating superpower while interacting with normal energy;

instead, dark energy transformed itself into a dark matter inhabiter that is embodied within a normal matter computer brain.

The inhabiter's imprint strings give rise to the universe's appearance of solidity. They are vibrating in a world whose geometry is shaped by the intertwined[265] synapses of billions of neurons and where dimensions are created by the warps and curvatures[266] that make up the contours of the normal matter computer brain. With these variables in place, the inhabiter and the normal matter computer brain are each given the opportunity to react and respond according to their nature and in such a way that balances or unbalances the equation of interaction. When the normal matter computer brain is directing itself and the dark matter inhabiter chooses the psyche self-centered, then the inhabiter becomes immersed in the world of normal matter. Immersion is something that happens[267] to an inhabiter embodied within a normal matter computer brain and was expected to occur by the pure awareness of the universe's dark energy. However, when the inhabiter chooses the psyche altruistic and the computer brain receives enough signal from the inhabiter's enabled non-ego-clinging imprint strings, then the inhabiter guides normal matter according to its dark energy compassionate nature.

The computer brain is unaware of the interaction it is having with the dark matter inhabiter; accordingly, as the equation of interaction gets balanced or unbalanced on a momentary basis, normal matter never cognitively suffers. During the interaction with normal matter, it is only the dark matter inhabiter that will suffer as it struggles to think with its cognitive states and experiences loss, aging, and death. Dark energy is the only energy in the universe capable of making a heroic gesture of love that is not impermanent or subject to change. Fully accepting the risks to itself and motivated by its desire to live in the universe harmoniously with normal energy, a focal point of dark matter embarked on a journey of discovery—but not without a safety net. Dark energy designed each inhabiter's configuration to include a focal point of itself. Therefore, each inhabiter will have a base of support in the form of its primordial consciousness.

While inhabiting a normal matter computer brain that is configured to produce data that enabled imprint strings create into a human psyche, the claustra's meta-algorithm and the inhabiter's focal point of dark energy with imprint strings are not directly connected. Instead, the meta-algorithm and the inhabiter's focal point of dark energy occupy two separate and distinct regions within the normal matter computer brain. The meta-algorithm and the inhabiter will interact by transmitting information to each other as signals, and part of the signals received from the computer brain by enabled imprint strings include neurotransmitter data.

Neurotransmitters are the computer brain's communicative programs and the means by which neurons are able to transfer information as a nerve impulse to another nerve fiber, a normal matter muscle fiber, and other normal matter structures. Although the computer brain has many different kinds of neurotransmitters, they work together to cause a cascade of events, allowing the computer brain to produce and process data while functioning as a unit. As neurotransmitters are released throughout the computer brain, where they land produces an immediate and effective reciprocation from normal matter's circuitry.

The inhabiter's imprint strings function similarly to the computer brain's neurons in this way: they are communicative among one another, and once enabled, they will produce information. Although unlike neurons that produce electrical patterns, enabled imprint strings produce information as electromagnetic waves representing transformed neural synchrony. Enabled imprint strings, until disabled, will influence the inhabiter's perception with electromagnetic waves of sense impressions and mental events. Enabled ego-clinging imprint strings create the psyche self-centered, and enabled non-ego-clinging imprint strings create the psyche altruistic via four steps and during two consecutive rotations of these steps. Two rotations of the four steps create one of the inhabiter's cognitive cycles.

- Step one: Enabled imprint strings receive signals sent by the normal matter computer brain that contain information that includes details about the computer brain's neurons, structures, and neurotransmitters involved in the creation of particular data.

The first rotation: Enabled imprint strings will select the data that pertains specifically to the normal matter computer brain to create sense impressions of gross consciousness.

The second rotation: Enabled imprint strings selecting the information received by normal matter from the network of interconnected computer brains spread throughout the universe to construct mental events present within subtle consciousness.

- Step two: During the first and second rotation, enabled imprint strings, having received the computer brain's signals, transform them into electromagnetic waves. Therefore, electromagnetic waves contain information that pertains, but is not limited to, neurotransmitter programs involved in the creation of particular computer brain data. Enabled ego-clinging imprint strings oscillate in way that is directed by the claustra's meta-algorithm, while enabled non-ego-clinging imprint strings oscillate in a way that is similar to enabled ego-clinging imprints, until they tap into the focal point of dark energy.

At the completion of step two, the electromagnetic radiation of both psyches travels across the surface of the focal point of dark energy and is moved by the kinetic energy produced by oscillating enabled imprint strings. Enabled imprint strings oscillate a specific way when moving the electromagnetic radiation from place to place on the surface of the focal point of dark energy, and this differs depending on the first or second rotation.

First rotation: Once the electromagnetic radiation reaches the interior surface of the focal point of dark energy, enabled imprint strings continue to oscillate but not in a way that moves the electromagnetic radiation—which means enabled imprint strings oscillate but temporarily cease to do so in a way that moves electromagnetic radiation or produces kinetic energy. Without kinetic energy produced by enabled imprint strings, the surface of the focal point of dark energy is temporarily smooth. This allows the primordial consciousness of the dark energy focal point to have a clear vantage point by which to analyze the information contained within electromagnetic waves. As the surface of the focal point is smooth, the view is not obscured by the rippling effect produced by kinetic energy.

Second rotation: Again, the electromagnetic radiation of both psyches travels in a similar fashion as it did during the first rotation, until it reaches the interior surface of the focal point of dark energy. During the second rotation, the oscillations of enabled imprint strings does not temporarily cease; instead, the oscillations of enabled imprint strings increase, and this causes the electromagnetic radiation of two psyches to coil upon itself to produce layers. The coiling of electromagnetic radiation creates what could be conceived as a funnel shape of layered electromagnetic radiation. Enabled imprint strings' continued oscillations move electromagnetic radiation of the two psyches so they layer but also causes the funnel-shaped electromagnetic radiation to levitate just above the surface of the focal point of dark energy.

The layering of the electromagnetic radiation of both psyches creates levels of electromagnetic waves that contain transformed data. When the electromagnetic waves are illuminated by the primordial consciousness of the focal point of dark energy, they produce the inhabiter's conscious experience. The funnel-shaped electromagnetic radiation has a narrowed base that widens with increasing layers of electromagnetic radiation. The base of the layered electromagnetic radiation contains the least amount of electromagnetic waves and represents the gross conscious level. As the layered electromagnetic radiation gets higher, it widens and contains more electromagnetic waves representing the subtle conscious level. The gross conscious level contains electromagnetic waves of sense impressions, and the subtle conscious level contains electromagnetic waves of mental events. The electromagnetic waves within the gross and subtle conscious levels contain electromagnetic waves with incorporated transformed neurotransmitter program data. Because the narrowed base of the funnel-shaped electromagnetic radiation does not directly make contact with the surface of the focal point of dark energy but levitates just above it, a space is created between the focal point's surface and layered electromagnetic radiation. This space represents the extremely subtle conscious level that is devoid of electromagnetic waves of sense impressions and mental events.

- Step three: During the first and second rotation of the four-step sequence, the inhabiter uses its primordial consciousness derived from its focal point of dark energy and is aware of the electromagnetic waves of both the psyche self-centered and the psyche altruistic. The inhabiter's primordial consciousness has abilities that exceed that of the computer brain and those of created conscious states, which means regardless of how many or what kind of neurotransmitters are released by the computer brain at any given moment, when transformed by enabled imprint strings and incorporated into electromagnetic waves, there is not one processing event that escapes the awareness of the inhabiter's focal point of dark energy.

Within the focal point of dark energy, primordial consciousness is illuminated in all directions, but this intelligence also emanates from the focal point. The amount of primordial intelligence that emanates from the focal point of dark energy does not change; whether illuminating the surface of the focal point of dark energy, extremely subtle conscious level or illuminating the gross or subtle conscious levels of the layered electromagnetic radiation, the amount is the same. Notwithstanding, amid step three during the second rotation of the four-step sequence, the electromagnetic radiation of both psyches is layered; thus, for primordial consciousness to illuminate the subtle conscious level, it must penetrate the layered electromagnetic radiation. Which means that unlike step three during the first rotation, primordial consciousness must do more than just illuminate the surface of the focal point of dark energy. What allows the inhabiter's primordial consciousness to permeate the layered electromagnetic radiation in step three during the second rotation is the increased oscillations of enabled imprint strings and the kinetic energy produced. Therefore, there is kinetic energy on the surface of the focal point, except for a centralized area located beneath the base of the layered electromagnetic radiation where there is no kinetic energy present. Thus, a pattern is produced, and to envision this, imagine the surface of the focal point of dark energy as being the iris of an eye. The centralized area devoid of kinetic energy is round and would be the pupil located centrally within the iris. There is a functional purpose to the iris/pupil pattern: where kinetic energy is present on the surface of the focal point of dark energy, the emission of primordial consciousness is blocked. However, primordial consciousness freely flows from the centralized area devoid of kinetic energy. Thus the emission of primordial consciousness is blocked in some directions but freely flows from the centralized area just beneath the levitating electromagnetic radiation. This yields a phenomenon in the form of a beam of primordial consciousness. The beam arises from the focal point and travels in an upward direction, passing through the layers of the electromagnetic radiation to the subtle conscious level.

Albeit, with each progressing level of layered electromagnetic radiation, there is more information for the inhabiter's primordial consciousness to illuminate. The inhabiter's beam of primordial consciousness is more narrowed at the surface of the focal point and spreads out at the subtle conscious level. Because of this, primordial consciousness is able to accommodate the increase in electromagnetic waves at the subtle conscious level.

The inhabiter's beam of primordial consciousness has within it a central region. This region represents a zone of intelligence with the capacity to move up and down to different levels within the layered electromagnetic radiation and in many different directions but in a way that differs from the beam itself. The beam illuminates the electromagnetic waves of both psyches simultaneously, which includes all the layers of the layered electromagnetic radiation up to the subtle conscious level.

The vividness of primordial consciousness does not change in the central region; rather, it is coherent intelligence that functions like a laser of awareness. The movement of the central region of the beam is governed by primordial consciousness's detection of transformed neurotransmitter data incorporated into electromagnetic waves. Neurotransmitter data that has been transformed by enabled imprint strings acts as a signal within electromagnetic waves that draws the attention of primordial consciousness and determines its reaction. However, the inhabiter's primordial reaction is not always the same as it relates to the beam, central region, and whether the transformed neurotransmitter data is detected by the focal point of dark energy in the electromagnetic waves of the psyche self-centered or the psyche altruistic.

When transformed neurotransmitter data is detected by primordial consciousness with its beam, it reacts by illuminating electromagnetic waves of both psyches simultaneously, which includes the gross and subtle levels within the electromagnetic radiation. When transformed neurotransmitter data is detected by primordial consciousness with the central region of its beam, this intelligence will follow where associated processing leads and for one psyche at a time. The central region of the beam of primordial consciousness will follow associated processing and react by illuminating the electromagnetic waves created by the imprint string type most enabled before repeating this process for the other type of enabled imprint string. Whichever string type is most enabled will be first, and primordial consciousness will use the central region of its beam sequentially, not simultaneously, to follow transformed neurotransmitter data. Primordial consciousness reacts by illuminating electromagnetic waves created during one complete cognitive cycle (two rotations of the four steps).

Although the universe's dark energy substrate layer is omnipotent and consistently stable and governs itself, when this same dark energy is a focal point within an inhabiter's configuration, its primordial reactions will be both specific as well as generalized. This disparity in primordial reactions is actualized by the dark energy focal point via its beam that has within it a central region, with the capacity to move in a way that differs from the beam itself.

The central region of the beam can be conceptualized two-dimensionally by the image of an arrow and its actions as being similar to a laser pointer. The tip of the two-dimensional arrow determines at which level within the layered electromagnetic radiation the laser pointer moves around, following transformed neurotransmitter data. As the laser follows transformed neurotransmitter data within electromagnetic waves of one psyche created during one cognitive cycle, coherent intelligence moves to the different levels within the layered electromagnetic radiation while following where associated processing leads. The central region represents primordial consciousness illuminating the electromagnetic waves of sense impressions and mental events present within one psyche, and its reactions therefore are unique. However, the beam is illuminating the electromagnetic waves of both psyches simultaneously; therefore, the reactions of primordial consciousness are undefined. Thus, the composition of the beam results in a phenomenon where both specific and generalized primordial reactions are actively present while the beam permeates the layered electromagnetic radiation.

As the inhabiter's data transformers, enabled imprint strings are the symbolic builders of doors that when opened create the inhabiter cognitive states. The focal point of dark energy is the figurative key maker and with primordial consciousness possesses all the keys to the doors created by enabled imprint strings. The central region of the beam and the beam itself is primordial consciousness that illuminates electromagnetic waves not with light, rather, with natural intelligence and thereby opens the many dimensions to the inhabiter's cognitive experience. The inhabiter's cognition emerges as a flow of mental experience when the inhabiter thinks with created cognitive states.

- Step four: Both the first and second rotations are distinct in that the inhabiter thinks differently.

Consequently, unlike when the inhabiter detects transformed neurotransmitter data with its beam and central region of primordial consciousness, while thinking with created conscious states, the inhabiter first becomes aware of transformed neurotransmitter data as the feeling tones of pleasant, unpleasant, or neutral.

First rotation: The inhabiter, in not thinking with a fully developed cognitive state, responds to the sense impressions of gross consciousness by predominately reacting, once aware of transformed neurotransmitter data as the feeling tones of pleasant, unpleasant, or neutral, versus cognizing its choices. A reverberation is produced that begins on the interior surface of the focal point of dark energy and travels to the peripheral surface. When the force of the reverberation reaches the peripheral surface of the dark energy focal point, this signals enabled imprint strings to use electromagnetic waves of sense impressions as building blocks to create mental events. The enabled imprint strings oscillate in a way so that kinetic energy is produced on the surface of the focal point of dark energy that moves the electromagnetic radiation of both psyches from the interior surface of the focal point to the peripheral surface where the imprint strings are located.

Second rotation: The inhabiter experiences physical sensations and emotions, thinks with thoughts, and has memories consisting of three-dimensional images or moving holograms. The inhabiter perceives electromagnetic waves as sense impressions and mental events that begin in step four, during the second rotation, with awareness of transformed neurotransmitter data as feeling tones. Feeling tones of pleasant, unpleasant, and neutral provide the inhabiter with an immediate and affective experience as the feeling that precedes a sense impression and a mental event. In this way, transformed neurotransmitter data serves to be a *mental bond* that tethers the inhabiter's primordial reaction to the inhabiter's reaction using its created cognitive states. Alas, there are multiple tethers because of the composition of the beam of primordial consciousness and the fact that the inhabiter will think with the cognitive states of both the psyche self-centered and the psyche altruistic, which create two contrasting cognitive experiences. The inhabiter using its created cognitive states to think will react to feeling tones, and those of pleasant and unpleasant will motivate an immersed inhabiter to choose the psyche self-centered. Whereas an inhabiter that understands feeling tones to be transformed neurotransmitter data and not a sense impression or a mental event does not use feelings to motivate it but instead desires to transform itself for benefit and thereby chooses the psyche altruistic.

Although the computer brain has a large repertoire of neurotransmitter programs, those created by the brain stem are particularly important because this structure sets the associative rules by which the meta-algorithm follows when processing information:

Acetylcholine: A neurotransmitter that when transformed by enabled imprint strings, the inhabiter will experience either a pleasant, unpleasant, or neutral feeling that precedes a sense impression or mental event that involves memory, sleep, muscle activation, and other nervous-system functions.[268]

Dopamine: A neurotransmitter that the normal matter computer brain releases when computing for reward. When transformed by enabled imprint strings, the inhabiter will experience it as a pleasant, unpleasant, or neutral feeling that precedes a sense impression or mental event.[269]

Norepinephrine (NE) or noradrenalin: A neurotransmitter that when released by the computer brain and transformed by enabled imprint stings the inhabiter will experience a pleasant, unpleasant, or neutral feeling that precedes a sense impression or mental event that involves arousal, focused attention, increased restlessness, anxiety, heart rate, and blood pressure; it also triggers the release of glucose from energy stores and increases blood flow to the skeletal muscles of the I character.[270]

Serotonin: A neurotransmitter of the normal computer brain that the inhabiter will commonly experience as a pleasant feeling that precedes a physical sensation or mental event; an immersed inhabiter cognizes that the mental event or sense impression is the cause of its well-being and happiness.[271]

Glutamate: This neurotransmitter program is included on the list because of its prominence throughout the computer brain, and when released, it enhances the synchronicity of the pulse propagating between two neurons.[272] When accumulated in a synaptic cleft and it spills over to adjacent synapses in other circuits, cross-talking between neurons occurs.[273] When transformed by enabled imprint strings, the inhabiter will experience a pleasant, unpleasant, or neutral feeling that precedes a sense impression and mental event, with an increase in volume[30]of the computer brain's data created by neuronal impulses throughout normal matter's circuitry.

Admittedly, the inhabiter's primordial consciousness possesses an astounding capacity to detect transformed neurotransmitter data within electromagnetic waves, albeit the inhabiter has limited ability in using its cognitive states to think. The inhabiter must use either the cognitive states within psyche self-centered or psyche altruistic, not both simultaneously, to think and cannot track the four steps used to create its cognitive states. The inhabiter perceives all four steps as a continuous flow of cognitive experience. Depending on how many electromagnetic waves with transformed neurotransmitter data are detected and illuminated specifically and generally by the inhabiter using its primordial consciousness determines how widely or narrowly this intelligence is disseminated. The inhabiter using its created cognitive states to think will experience this discrepancy: if there is an abundance of electromagnetic waves illuminated, then primordial awareness is being disseminated in multiple directions. Accordingly, the inhabiter experiences less lucid cognitive states with a decreased attention span. The more electromagnetic waves within one cognitive cycle that are illuminated by primordial consciousness with the central region of its beam of awareness, the less distinct the details of a metal event and less vivid a sense impression. Contrastingly, if there is a minimal number of electromagnetic waves illuminated, then primordial consciousness is being disseminated narrowly and in fewer directions. The fewer electromagnetic waves that are illuminated by primordial consciousness using the central region of its beam of awareness, the more distinct the details of the metal event and more vivid the sense impression. Correspondingly, the inhabiter experiences this as its created cognitive states becoming more lucid, and the inhabiter's attention span is lengthened.

An example of this phenomenon uses the two-dimensional arrow image and laser pointer previously described: Imagine an arrow stretching from the surface of the focal point of dark energy that is encompassed within a beam of light. The beam of light represents primordial consciousness generally illuminating electromagnetic waves of both psyches simultaneously up to the subtle consciousness level. The arrow represents the central region of the beam of light and is coherent primordial consciousness specifically illuminating electromagnetic waves of one psyche while following transformed neurotransmitter data. In this example, the tip of the arrow is located at the subtle conscious level and is moving like a laser pointer within the beam of light at this level, while illuminating electromagnetic waves wherever processing leads. The inhabiter using its created cognitive states to think is unaware of what its primordial consciousness is doing. Instead, the inhabiter is aware of the three-dimensional hologram of a mental event and therefore is using its created cognitive states to think. This mental event would be the primary focus that the inhabiter perceives with the cognitive states of one psyche. However, due to the conformation of the beam, though the tip of the arrow is located at the subtle conscious level and acts like a laser pointer, the electromagnetic waves at the gross level are also being generally illuminated by the beam of light. Thus, within the inhabiter's peripheral awareness, there will be a secondary focus. The primary mental event might be broadly perceived by the inhabiter as a room or narrowly perceived as an object within the room, depending on how many electromagnetic waves were illuminated at the subtle conscious level. The concurrent secondary focus might consist of the inhabiter's vague awareness of the inspiration or expiration of the breath or a beating of the heart. If the tip of the arrow gets changed and moves to the gross conscious level, then in this situation the breath or beat of the heart becomes the primary focus, but in the periphery of the inhabiter's awareness, there will be a vague awareness of the ambiance of the room. If the central region of the beam vacillates at the gross conscious level while illuminating electromagnetic waves, then in this scenario, from the inhabiter's viewpoint, the breath might be the primary focus and the beating of the heart a secondary focus, but in the periphery of the inhabiter's awareness, there will be a vague awareness of the ambiance of the room.

These examples are simplified, as the factors that culminate to produce the dimensions of the inhabiter's cognitive experience are highly complex and include the following: the way data was processed by the computer brain and the number of neurotransmitters released; the transformation of the computer brain's data by enabled imprint strings into electromagnetic waves, the inhabiter's generalized and specific primordial reactions; while thinking with cognitive states, the inhabiter's reaction to feeling tones and its quick or thoughtful response in choosing the psyche by which to experience momentary reality.

However, if the tip of the arrow further descends and moves to the space between the surface of the focal point and the layered electromagnetic radiation, then it is located at the extremely subtle conscious level. The space that is the extremely subtle conscious level is created by the layered electromagnetic radiation levitating above the surface of the focal point. As the beam of light widens the farther it travels from the focal point, near the surface of the focal point it is narrowed. Thus, the extremely subtle conscious level is illuminated by a narrowed beam of light with a coherent arrow of primordial consciousness. However, there are no electromagnetic waves with incorporated transformed neurotransmitter data present in this space. That means this space is devoid of the signals that draw the attention of primordial consciousness and determine its reaction, and there is nothing governing the movement of the arrow. The central region of the beam of primordial consciousness (arrow) is not illuminating data of sense impressions or mental events at the extremely subtle conscious level, thus the inhabiter cannot think with its created conscious states. The central region of the beam of primordial consciousness is illuminating the space itself at the extremely subtle conscious level. In this way, the space at the extremely subtle conscious level serves to be a *mental bond* that tethers the inhabiter's primordial reaction to the inhabiter's reaction that is unsupported by sense impressions or mental events. Due to the conformation of the beam, though the central region of the beam is located at the extremely subtle conscious level and the space itself is the primary focus, the electromagnetic waves at the gross and subtle conscious levels are also being generally illuminated by the beam to create a vague secondary focus. This phenomenon could be summarized as *awareness of awareness*. Although the inhabiter is still on its journey, interacting with normal matter while embodied within a normal matter computer brain, when the tip of the arrow (central region of the beam) is located at the extremely subtle conscious level, the inhabiter is nearly, though not exactly, experiencing how it resides blissfully as the dark energy substrate layer in the nonseparable universe.

The concept of time is the reference by which the inhabiter contextualizes its mental experience. Time commences when the primordial beam of awareness emanates from the surface of the focal point and reaches the subtle conscious level and when the central region of the beam illuminates the electromagnetic waves of a psyche. The inhabiter cognizes time and experiences it in many different ways, such as waking from sleeping or through the perception of the I character aging. Depending on whether the computer brain is processing data for the core-I character or autobiographical-I character and where the central region of the beam of primordial consciousness is located, time will be experienced by the inhabiter as present moment, past tense, or as future events. The electromagnetic waves at the gross conscious level represent the early stages in the development of the inhabiter's cognitive states. Thus, when the central region of the beam of primordial consciousness is located at the gross conscious level, the inhabiter will have a limited awareness of time. When the central region of the beam of primordial consciousness is illuminating the extremely subtle conscious level, the inhabiter will not experience sense impressions or perceive mental events or conceptualize the concept of time.

The fact that the inhabiter has limited abilities and must think using the cognitive states of the psyche self-centered or psyche altruistic at any given moment was not a happenstance occurrence; rather, this was dark energy's method of balancing the equation of its interaction with normal matter and adapting to its life of conditioned existence via simplification. The universe's dark energy substrate layer does not think with created cognitive states or electrical patterns, but transformed as a dark matter inhabiter, dark energy must adapt to thinking with created cognitive states. When designing the interaction with normal matter, dark energy could have avoided its transformation to dark matter and just been a focal point of dark energy. However, because dark energy is imbued with tremendous power with the potential to overwhelm the normal matter computer brain, if dark energy did not channel its power in some way, the interaction could hardly be construed as being meaningful. Instead, this would equate to dark energy exhibiting an unfair influence and use of its power. Hence, dark energy configured itself to be the dark matter inhabiter with a focal point of dark energy where primordial consciousness plays an adaptive, not a domineering, role in the interaction with normal matter. When the inhabiter uses its primordial consciousness derived from the focal point of dark energy to illuminate electromagnetic waves, the electrical pattern language of the normal matter computer brain becomes decipherable to the inhabiter while thinking with created cognitive states. The balance of power between the inhabiter and the computer brain is further equalized by the fact that enabled imprint strings process information at the speed that is comparable to the computer brain's neurons. The computer brain communicates with itself through billions of neurons firing randomly and independently. Enabled imprint strings work for the inhabiter by functioning as data transformers[274] that decode the colossal amount of information generated by the normal matter computer brain. However, while embodied within a computer brain, the dark matter inhabiter must learn to adjust to thinking with created cognitive states quickly, as the longer it takes the inhabiter to do this, the more this unfairly impacts the normal matter computer brain by prolonging the interaction. Normal matter was drawn to engagement by the dark matter inhabiter, and unable to escape the interaction, normal matter configured as a computer brain remains trapped until inhabiters make their informed decisions.[275] As there are dozens of sense impressions and mental events created in a single second, it means that there are dozens of rotations of the four steps in a single second. To appreciate how complex this situation is, imagine each sense impression and mental event as being the spices and ingredients of two lavish soups. One soup is called psyche self-entered that when chosen weakens the inhabiter through inaccurate perception. The other soup is called psyche altruistic that when chosen nourishes the inhabiter through accurate perception and discerning recognition. If the inhabiter had to master identifying every spice and ingredient in the lavish soups before having the ability to determine which soup would result in its deprivation or nourishment, the inhabiter might starve before it could make a beneficial choice. Therefore, in order to enhance the inhabiter's ability to learn quickly and adjust to thinking with its created cognitive states, simplicity was a design detail dark energy incorporated into the equation of interaction. The inhabiter does not need to discern all the ingredients and spices of the soups to make a beneficial choice. Rather, the inhabiter simply needs to evaluate its motivations and discern the I character's words and actions to determine which psyche it is perceiving. Thus in one taste consisting of motivation and the words and actions of the I character, the inhabiter has clues and the ability to determine which psyche it is perceiving. To nourish itself, the inhabiter must choose the psyche altruistic.

What is real? When the inhabiter chooses one of the psyches to experience its momentary reality, the cognitive states of that psyche define situations as real. As the inhabiter's choices enable and disable its imprint strings that send signals to the computer brain and result in alternative programming or allow the computer brain to process data according to its nature, the situations perceived by the inhabiter are real in their consequence. Psyche self-centered and psyche altruistic contain cognitive states with disparate thoughts, motivations, and beliefs. Feelings tones of pleasant, unpleasant, and neutral provide the inhabiter with an immediate and effective experience. However, feeling tones are not emotions, memories, or thoughts that allow the inhabiter to formulate an objective opinion. Correspondingly, the inhabiter's interpretation of a situation within a psyche is not objective but a subjective perception of the situation. The inhabiter will experience its subjective impressions through the body of the I character and by interpreting situations involving the character. Because the sense impression and mental event exist only within the psyche, neither has an actual reality. Therefore, an objectively correct interpretation of whatever emerges within a particular psyche is not important for the purposes of helping the inhabiter make beneficial choices. What matters is that the psyche chosen by the inhabiter will influence the present situation, as enabled imprint strings do not simply disappear or disable themselves and act as promissory notes with regard to the inhabiter's future happiness or suffering. If the inhabiter chooses the psyche self-centered, then the inhabiter is experiencing life in a way that perpetuates its delusion because the inhabiter is misapprehending illusory reality that it believes is real.[276] The fact that the inhabiter is interacting with the normal matter computer brain but also is connected and communicating with other inhabiters via the phenomenon of entanglement, the nonseparable universe is affected by the present situation in many complex ways.

The inhabiter has the opportunity to weigh its options before responding and choosing the psyche by which to experience momentary reality. However, the inhabiter can forfeit this option by quickly responding before perceiving the alternative response within the cognitive states of the other psyche. Notwithstanding, whether the inhabiter makes a careful or hasty response, having made a choice, the electromagnetic radiation of both psyches dissolves into the focal point of dark energy, and this causes a reverberation with a force that correlates directly with the inhabiter's motivation. The reverberation moves across the surface of the inhabiter's focal point of dark energy until it reaches the enabled imprint strings and will determine the type of imprint strings and how many of them will be disabled. As the strings are arranged on the inhabiter's focal point of dark energy as ego-clinging and non-ego-clinging pairs, a change in a magnetic charge through one type of enabled imprint string causes a magnetomotive force across the other that disables or enables the other type of imprint string. Specifically, an ego-clinging imprint string disables a non-ego-clinging imprint string, and the opposite mechanism is at work when non-ego-clinging imprint strings are enabled. The enabled imprint strings that are left will transmit alternating signals of magnetic charges, which will be picked up by the normal matter computer brain's antenna of the brain stem and spinal cord and will be received by the meta-algorithm in the interclaustral pathway.

Embodied within a normal matter computer brain, the inhabiter experiences a psyche that deceptively displays what the inhabiter understands to be its life. In fact it would be more accurate if the inhabiter adopted the viewpoint that each psyche is analogous to a normal matter computer brain / enabled-imprint-string-based simulated environment. Within the psyches interacting energies are made to appear solid, and where there is unpredictability, predictable order appears. Despite that the psyches are an illusory representation of reality, presented in this way, the complexities of reality abate. The universe's dark energy substrate layer created imprint strings so that each dark matter inhabiter has the opportunity to make the most of its experience while interacting with normal matter via the psyches. When electromagnetic waves are illuminated by the inhabiter's primordial dark energy consciousness, each inhabiter is given the gift of perception with the mechanics of a three-dimensional hologram and aesthetics of a life that engage the inhabiter. The psyches also motivate the inhabiter's responses that promote learning and teach the inhabiter to solve its problems[277] because the psyches contain the inhabiter's created cognition that includes emotion and specific thoughts.[278] It is as though the inhabiter is participating in a game, the game of life. Albeit, life when apprehended by the inhabiter to be an interactive game called the game of life is not meant to minimize the significance of conditioned existence or imply that life is trivial. Indeed, the interaction between the dark matter inhabiter and the normal matter computer brain is meaningful and the consequences significant when considering that the inhabiter's responses and the choosing of a psyche is tallied by enabled and disabled imprint strings. When ego-clinging imprint strings are abundantly enabled, the inhabiter will suffer while participating in the game of life, whereas when non-ego-clinging imprint strings are richly enabled, the inhabiter has transformed itself into a skilled participant in its interaction with the normal matter computer brain.

The game of life includes accessories that allow the inhabiter to participate in the game. The I character is the avatar of interaction between the inhabiter and the normal matter computer brain and is named the pawn. The pawn allows the inhabiter to participate in the game of life at the subtle conscious level of the layered electromagnetic radiation. However, to the computer brain, the I character is a metastable construct made of neural synchrony that the meta-algorithm uses to link agents and by doing so is able to load program data. When this data is received as signals and transformed by enabled imprint strings to electromagnetic waves with incorporated transformed neurotransmitter data and illuminated by the central region of the beam of primordial consciousness at the subtle conscious level, the inhabiter perceives the pawn's appearance and every aspect of the microcosm that revolves around this singularity. Yet no version of the I character intrinsically possesses any solidity, form, shape, or color.[279] The inhabiter fabricates this character with its created cognitive states and is a mental projection[280] within the subtle conscious level of the layered electromagnetic radiation. The appearance of one I character with a body that moves and speaks in the three-dimensional hologram of a psyche is an illusion, as there are many versions of the pawn. At each moment, the meta-algorithm imperceptibly vacillates between computing for the core-I and autobiographical-I character. The meta-algorithm must maintain internal states for the computer brain to continue to function as a unit and will correlate synchrony of biologic information first. [281] The proto-I character is the primitive and internal milieu presence programming within all versions of the I character.[282] From this foundation, the inhabiter's created conscious states proceed with a commanding urge for survival and to maintain stability. Therefore, the inhabiter's sense of self emerges with a specific psychological experience[283] perhaps five hundred milliseconds[284] after electromagnetic waves are illuminated by primordial consciousness. The Inhabiter becomes aware of a pleasant, unpleasant or neutral feeling followed by a sense impression and then with mental events the inhabiter begins identifying itself as the pawn. To compute for the here and now, the meta-algorithm uses the core-I[285] character, and when historical data is accessed, the autobiographic-I[286] character emerges. Each character fulfills a specific programming purpose, allowing normal matter configured as a computer brain to function as a unit. Albeit, from the inhabiter's viewpoint at the subtle conscious level, its life is perceived through the eyes of the pawn, yet neither the normal matter computer brain nor the dark matter inhabiter has eyes. The normal matter computer brain has circuitry, with each optic nerve containing between 770,000 and 1.7 million nerve fibers[287] that relay information to neural nets within the computer brain's occipital lobe, thalamus, and hypothalamus. This information will be processed and combined with data generated by the brain stem, where the brain stem's interconnection to the implicit layer of hybrid architecture forms a recursive pattern of circuitry.[288] The meta-algorithm uses this information to serially sequence agents that enabled imprint strings transform into the electromagnetic waves that when illuminated by the central region of the inhabiter's beam of primordial consciousness at the subtle conscious level, the inhabiter will perceive parts of the pawn's body.[289] The inhabiter's reality is symbolically represented in the psyches and reveals the relative and complete dimensions of conditioned existence. The eyes of the pawn are an example of a simplified relative symbolic representation, and a deeper symbolic representation would be the iris/pupil pattern (previously described): kinetic energy present on the surface of the inhabiter's focal point of dark energy blocks the emission of its primordial consciousness, but primordial consciousness freely flows from the centralized area devoid of kinetic energy.

Life embodied within a normal matter computer brain is complex and unpredictability is the norm that is a fact that is not initially grasped by an inhabiter as it begins its odyssey of discovery. Instead, the degree to which an inhabiter's ego-clinging imprint strings are enabled directly correlates with its perception that the pawn is special and that somehow the rules of life are for other characters and do not apply to the I character that the inhabiter perceives as itself.[290] Within the psyches, there are characters the inhabiter perceives as loved ones, strangers, nonhuman life forms, and enemies. Although these illusory characters exist only within the psyche, they do symbolically represent the I characters of other inhabiters that are spread throughout the universe. By virtue of the phenomenon of entanglement, each inhabiter's focal point of dark energy receives quantum signals from the universe's dark energy substrate layer with information about another inhabiter's I character and will be incorporated contextually into both psyches in a way that will make sense to the inhabiter because it complements mental events. This means an inhabiter may perceive its pawn a certain way that only slightly corresponds with the way this character is perceived by another inhabiter within their psyches. In fact, the inhabiter cannot be certain that the words and actions of its pawn within psyche self-centered or psyche altruistic are actually occurring within the psyches of another inhabiter while both inhabiters are entangled. When guiding itself, the normal matter computer brain is unpredictable as to the exact data it will generate with its circuitry and how the meta-algorithm will process information. When the inhabiter's focal point of dark energy receives quantum signals from the universe's dark energy substrate layer and enabled imprint strings assimilate this information into electromagnetic waves and later send signals to the computer brain, the communication between inhabiters is transformed to neural synchrony. Thus, the inhabiter's communication with another inhabiter becomes a part of the normal matter computer brain's history. While the meta-algorithm does not see inherent meaning, it does react to degrees of neural synchrony that it modulates to produce agents. Some of these agents when linked by the meta-algorithm and sent as signals are transformed by enabled imprint strings to electromagnetic waves with incorporated neurotransmitter data. Thus, the original quantum signal sent by the universe's dark energy substrate layer that was received by an inhabiter's focal point of dark energy can be dramatically altered and might be represented within the psyches very differently. Sometimes when electromagnetic waves are illuminated by the inhabiter's beam of primordial consciousness with a central region, they bring into existence within the psyches menacing characters, phenomenal disasters, and intrusively disturbing scenes. However dangerous or upsetting the inhabiter perceives these manifestations to be, in actuality they are not caused by an independent entity or force external to its focal point of dark energy with a peripheral surface of enabled imprint strings. The characters seen doing harm and the harm itself are totally devoid of any inherent existence.[291] This is the complexity of a universe of inhabiters witnessing psyches with illusory worlds while embodied within normal matter computer brains and underlies the illusion of a multiuniverse. Apart from these empty phenomena of enabled imprint strings receiving and transforming the computer brain's signals, the inhabiter's primordial reactions and cognitive choices, what is there for the inhabiter to gain or lose, to want or reject? An inhabiter that has transformed itself into a skillful participant in the game of life will consistently choose psyche altruistic and as a result will have an alternative perspective of difficult characters and upsetting scenes that are displayed within the psyches. These characters and situations are extremely valuable to the inhabiter on its odyssey of discovery while interacting with the normal matter computer brain, as the inhabiter can use them as measuring sticks of its ability to make constructive responses that disable its ego-clinging imprint strings.

The inhabiter, perceiving the cognitive states within the psyches, will formulate what the inhabiter believes to be memories of its life even though none of them actually occurred,[292] as they were inaccurately perceived. An inhabiter that has no recollection of its previous form as part of the universe's dark energy substrate layer will experience difficulty seeing beyond its present life[293] as the pawn. The viewpoint of an inhabiter immersed in created cognition of the psyche self-centered is that the pawn is an enduring self-perception that the inhabiter clings to as itself, even though, in fact, there is nothing to cling to.[294] But the inhabiter cherishes this character as itself nonetheless and gets used to its inventive perception until the I character seems to exist as a distinct entity.[295] This is the product of the inhabiter's enabled ego-clinging imprint strings and the feeling tones of pleasant and unpleasant exerting their influence on the inhabiter's perception, and the inhabiter begins to think in terms of *I, my,* and *mine*, becoming a slave to its inaccurate beliefs.[296] Thinking about "my" body, possessions, money, relatives, friends, and enemies[297] by the inhabiter's responses in consistently choosing the psyche self-centered, it diligently perpetuates the enabling of more of its ego-clinging imprint strings. Underlying each of its conscious states for reward or punishment, pleasure or pain, approach or withdrawal, and personal advantage or disadvantage,[298] the inhabiter will perceive the pawn in the psyche self-centered with words and actions that oppose those of the pawn in the psyche altruistic. As the inhabiter focuses on the cognitive states of the psyche self-centered, the pertinent point grasped by the inhabiter will be rewarding the pawn, striving to maintain relationships so that the I character benefits and eliminates threats.[299] However, if the inhabiter has responded by choosing this psyche by which to experience its momentary reality, then the inhabiter has gained and conquered nothing, only enabled its ego-clinging imprint strings.

The inhabiter's fundamental error in perception—an error that motivates its actions and is the catalyst for its unrewarding responses—is its misidentification of itself as the pawn. However, primordial reactions and responses influenced by feeling tones is what causes the inhabiter to make inaccurate perception the foothold for its choices. Ergo, when the inhabiter enables its ego-clinging imprint strings, the inhabiter will be unable to profoundly bend its perception concerning what it thought it knew about itself and the phenomenal world. Enabled ego-clinging imprint strings will never generate the electromagnetic waves that when illuminated by primordial consciousness will result in cognitive processing that produces the inhabiter's disbelief in an independently existing I character. When cognitively absorbed, the inhabiter will habitually choose the psyche self-centered and will have a mistaken perception of personal control as it identifies itself as the pawn.[300] The illusion of the pawn will be intrinsically reassuring and rewarding to the inhabiter because of the comfort derived from the mistaken perception of personal control.[301] What is comforting to the inhabiter is its perception of an identifiable presence in the illusory world of normal matter, as this serves as the reference point for the inhabiter to assert, "I am here and am important!"

The inhabiter's situation while embodied within a normal matter computer brain might be better understood with a financial analogy:

- Currency-enabled imprint string.
- Accumulation of wealth-enabling non-ego-clinging imprint strings.
- Accumulation of debt-enabling ego-clinging imprint strings.
- Bank—the peripheral surface of the inhabiter's focal point of dark energy where imprint strings are situated.
- The method by which the inhabiter monitors wealth/debt: at the subtle conscious level by monitoring the words and actions of the pawn and through motivations experienced by the inhabiter.
- Beneficial long-term financial plan: to complete the interaction with the computer brain in a way that benefits normal matter, the inhabiter, and the nonseparable universe. With a financial viewpoint, the inhabiter's mission is to transform itself into its own skillful financial adviser by discerning its motivations and to monitor the pawn. If the inhabiter determines that the concept of self is its primary motivation, then the inhabiter has experienced the influence of pleasant and unpleasant feeling tones. Accordingly, the inhabiter will feel attachment and aversion and will perceive the pawn's words as being self-serving and its actions steeped

119

in acquisition, avoidance, and aggression. The inhabiter in this situation is perceiving the psyche self-centered, and if the inhabiter chooses this nonlucid dream, then it has accumulated debt by enabling its ego-clinging imprint strings. However, if the inhabiter determines that what motivates it is to be of benefit and does not feel attachment to the pawn, then the inhabiter has experienced the influence of altruism. Consequently, the inhabiter will feel tapped into the allegorical financial landscape in a way where its future to be of benefit is secured while perceiving the pawn's words to be loving and benevolent with actions that are imbued with compassion. The inhabiter in this situation is perceiving the psyche altruistic, and if the inhabiter chooses this lucid dream, then the inhabiter has accumulated wealth and long-term financial growth by the disabling of its ego-clinging imprint strings.

Enabled non-ego-clinging imprint strings serve the purpose of enriching the inhabiter's interaction with normal matter, allowing the inhabiter to perceive and interpret the brain's electrical patterns but with clarity. The inhabiter, using its primordial consciousness and enabled non-ego-clinging imprint strings, will be challenged while altering the computer brain's sensorimotor processing and change the meta-algorithm's methods for sequencing agents of neural synchrony. Reprogramming the computer brain takes time, and if done too rapidly, the meta-algorithm may react as if it is being threatened. Synchrony must be accepted by the meta-algorithm, and if the process of reprogramming the computer brain's circuitry is done too drastically and an agent contains information that does not fit well within the computer brain's fabricated environment, the meta-algorithm will modify the timing of its processing to eliminate it.

The electromagnetic waves produced by enabled non-ego-clinging imprint strings include transformed neurotransmitter data that was altered by the inhabiter using the primordial consciousness of its dark energy focal point when these imprint strings tapped in. Therefore, when transformed neurotransmitter data is detected within electromagnetic waves of the psyche altruistic, primordial consciousness is reacting according to its nature, and its reactions are purposeful and compassionately predictable. Detecting transformed neurotransmitter data of its creation, primordial consciousness reacts by anchoring the central region of its beam at the gross conscious level as the means by which to stabilize its coherent awareness. In this way, the cognitive states of the psyche altruistic are tethered to primordial reactions so that the inhabiter has the means by which to sense its primordial presence, and the variable of compassion within the equation of interaction is extended to the inhabiter. The inhabiter must be willing to accept the changes in its relationship to the pawn, and this will not be easy for the inhabiter. Doing so means the inhabiter will have no identifiable living being by which to identify itself within the cognitive cycles of the psyche altruistic. The inhabiter's focal point of dark energy is the inhabiter's perfectly black body, not perceived by the computer brain because its circuitry cannot compute for color at very low intensities.[302] However, accurate information is powerful, and the inhabiter with a sufficient number of enabled non-ego-clinging imprint strings has the capacity to understand facts about its life of conditioned existence while embodied by a normal matter computer brain. When the central region of the beam of primordial consciousness is anchored and specifically illuminating electromagnetic waves of the psyche altruistic, in the background the beam is generally illuminating the layered electromagnetic radiation. Thus the inhabiter's cognizance of the primordial presence within its configuration is vivid and nonconceptual, and its viewpoint is expanded. The primordial presence within the cognitive states of the psyche altruistic might be perceived by the inhabiter as a primary focus or a secondary focus. The inhabiter's perception of the psyche altruistic might be, for example, the following:

Primordial presence using the example of the breath as a primary focus while the central region is anchored at the gross conscious level. The beam creates the secondary focus, such as ambient environment, while illuminating the layered electromagnetic radiation: an all-embracing vivid awareness of the sensation of breathing that is freed from the constraints that emerge when the inhabiter is conceptualizing the pawn, with a peripheral awareness of the ambient environment.

Primordial presence as a secondary focus such as the breath, the primary focus possibly using nature landscape. The central region is anchored at the subtle conscious level, and the beam is illuminating the layered electromagnetic radiation: a lucid and expansive awareness of the details of a nature landscape while in the periphery of awareness a nonconceptual sensation of breathing.

If the inhabiter is to understand the complexities surrounding its life of conditioned existence, its relative and infinite viewpoints must be connected. The relative dimensions of the inhabiter's understanding correspond to the inhabiter's conventional perception where the phenomenal world is understood as being made of individual particles.[303] When the inhabiter understands its relative perception to be the result of appearances, not external to the surface of its focal point of dark energy but rather produced by causes and conditions that have no independent or permanent existence, then inhabiter's conventional perception is connected to its infinite viewpoint.[304] Each inhabiter will be challenged in different ways, and each will respond in its unique way to the psyches created by enabled imprint strings. The inhabiter that uses its psyches to learn facts and analyze them so that its relative and absolute understanding of its existence are connected possesses an expanded viewpoint. Therefore, the inhabiter has the ability to stay on task with its long-term aim while training itself so that it learns what actions to avoid and adopt and calmly is able to witness its cognitive states.[305] This describes an inhabiter that is reconstructing itself from a passive viewer of the psyche self-centered to a mental detective that is looking for clues within its cognitive experience. As a mental detective, the inhabiter discerns its motivations and interprets the pawn's words and actions that clearly depict the psyche altruistic. Once detected, the inhabiter chooses this psyche as the means by which to experience momentary illusory reality. When the I character is perceived by the inhabiter not as "self" but rather a pawn, the inhabiter becomes a skillful participant in its life of conditioned existence while interacting with the normal matter computer brain.[306]

The inhabiter has the opportunity to weigh its options before choosing the psyche by which to experience momentary reality. If the inhabiter repeatedly chooses the psyche altruistic, more ego-clinging imprint strings will be disabled. This means the psyche altruistic will be the option that emerges before the psyche self-centered in the second rotation, during step four. In so much as the inhabiter has become adept in discernment and thereby increased its learning curve by knowing what content to look for, then a hasty choice in choosing a psyche can be favorable as long as the inhabiter picks the psyche altruistic. This is a figurative mental hack performed by the inhabiter that, having clearly recognized psyche altruistic, immediately chooses it and therefore does not perceive the psyche self-centered as an option. Because what the inhabiter does affects the normal matter computer brain, a cascade of constructive programming will ensue. The more sustained the signals received by the computer brain from enabled non-ego-clinging imprint strings, the faster circuitry gets reprogrammed because neurons that repeatedly fire together will wire together in circuitry.

The dark matter inhabiter and the normal matter computer brain are two very different forms of energy; therefore, *glitches* will arise while interacting. If there was only one type of imprint string enabled, the inhabiter would have less of a problem interacting with normal matter. Instead, the inhabiter has two types of imprint strings enabled to varying degrees and therefore two psyches that create cognitive states with opposing motivation and beliefs. The problems that arise between the inhabiter and the computer brain can be attributed in part to the fact that the inhabiter perceives meaning with its cognitive states. This leads the inhabiter to make inaccurate assumptions about the sense impression and mental events experienced in psyche self-centered that do not correspond with the reason the normal matter computer brain generated the information. Although the inhabiter may be a compliant spectator of the psyche self-centered, with no actual reality, this psyche represents the inhabiter having a nonlucid dream or a nonlucid nightmare. This means that sooner or later, the inhabiter is going to have to choose the psyche altruistic as its response to conquer its inaccurate beliefs. However, the content of this psyche also has no actual reality; thus, this psyche represents the inhabiter having a lucid dream or a lucid therapeutic nightmare.

What distinguishes the "Psyche Self Centered" experienced by the Inhabiter as a non-lucid dream from that of a non-lucid nightmare is the degree by which the Inhabiter unbalances the equation of interaction by its choice of this psyche. What influences the inhabiter and motivates it to choose the psyche self-centered nonlucid dream is the feeling tones of pleasant and unpleasant. Therefore, the inhabiter is driven by attachment and aversion. The inhabiter experiences this through desire and the need to perpetuate pleasant experiences while avoiding those experiences that are unpleasant. When the inhabiter chooses the psyche self-centered and enables its ego-clinging imprint strings, the inhabiter suffers, though it might not be aware of this. However, when the amount of suffering is so significant that it interferes with the inhabiter's ability to interact meaningful with the normal matter computer brain, then this is problematic and constitutes a nonlucid nightmare. These nonlucid nightmares are produced by the normal matter computer brain's circuitry, the population of enabled ego-clinging imprint strings, and the inhabiter's beam of primordial consciousness with a central region that detects transformed neurotransmitter data and follows where processing leads. There are many different flavors of problematic psyche self-centered nonlucid nightmares and ways in which the inhabiter's suffering interferes with its life of conditioned existence. The inhabiter suffers via unpleasant thoughts, memories, emotions, impulses, and urges and by experiencing painful physical sensations. While witnessing the psyche self-centered as a nonlucid nightmare, the inhabiter will perceive depressogenic situations, and the pawn will exhibit neutralizing or avoidance behaviors.

While embodied within a normal matter computer brain, there are two variables of the equation of interaction that all inhabiters are obliged to contend with: the configuration of normal matter as a computer brain, which impacts its ability to function as a unit, and uncertainty as related to the exact moment the computer brain computes for impermanence. Inhabiters will contend with these two variables in their own unique way, but the more unbalanced the equation, the greater the inhabiter's suffering. When the equation of interaction becomes critically unbalanced, these two variables conspire to produce the psyche self-centered nonlucid nightmare where the inhabiter may feel it is tumbling down a black hole so deep in despair that death appears to be the only escape route. All inhabiters are susceptible to this nightmare, yet the inhabiter, having chosen the psyche self-centered and having a nonlucid dream, will believe itself to be immune from suffering. Misguided self-assurance and unanchored security poses the greatest danger to an inhabiter that chooses the psyche self-centered and produces the inhabiter's belief that the pawn's ability to accumulate wealth, status, and power-connected relationships provides protection. The pawn may be the most powerful entity within the world of the psyche self-centered, but the fact remains that if the normal matter computer brain unexpectedly computes differently, the equation of the interaction can very quickly become critically unbalanced: the inhabiter will witness the pawn brought to its very knees perhaps by mental illness and depression of which no amount of money can buy a quick fix, or a way out, but can cause a tremendous amount of suffering that the inhabiter will experience.

What is similar between inhabiters is the method by which maladies of perception can be managed, and this is via the psyche altruistic, lucid therapeutic nightmare. When the inhabiter chooses the psyche altruistic because it wants to address its maladies of perception and get better, then this psyche is transformed to a lucid therapeutic nightmare. When the inhabiter chooses the psyche altruistic with this motivation, then its primordial consciousness and the cognitive states within this psyche will work synergistically. This counteracts the constrictive tether that exists between the inhabiter's primordial consciousness and cognitive states that produce the psyche self-centered nonlucid nightmare. This means the central region of the beam of the inhabiter's primordial consciousness reacts to transformed neurotransmitter data with the purpose of illuminating electromagnetic waves in a way that will transform the inhabiter's pivotal motivations and beliefs. At the subtle conscious level, the inhabiter using its created cognitive states thinks like a mental detective, and its task is the following: identify stress increasers (obsessions) and the pawn's neutralizing and avoidance behaviors (compulsions) that were problematically displayed in the psyche self-centered nonlucid nightmare.[307] The inhabiter will then recreate these via the psyche altruistic that is experienced by the inhabiter as a lucid therapeutic nightmare. Although there is no one correct action or recreation that fits for all inhabiters, psyche altruistic experienced by the inhabiter as a lucid therapeutic nightmare is highly effective when there is also a primordial presence. Within the cognitive states of the psyche altruistic, the primordial presence is commonly represented by the breath and will be a primary focus when the inhabiter is training itself in sustained calm cognitive states and a secondary focus when the inhabiter is lessoning its reactivity to fear and stress via graduated exposure. Exposure consists of the inhabiter recreating within the psyche altruistic problematic, life-interfering sense impressions or/and mental events of the psyche self-centered and experiencing them as a lucid therapeutic nightmare. This means with a primordial presence, the inhabiter is aware of transformed neurotransmitter data as the feeling tone, which may be unpleasant and precedes a physical sensation and/or a mental event such as a situation in thought and memory and may include the inhabiter experiencing intrusive impulses and urges. By rating the level of anxiety experienced from one to ten, with ten being the greatest, the inhabiter anchored with a primordial presence will witness and experience the problematic starting with those with a rating of four, five, or six for one hour, or until its anxiety is decreased by half. The pawn perceived within the psyche altruistic experienced by the inhabiter as a lucid therapeutic nightmare will be the version most suited to address the inhabiter's nonbeneficial responses.

Motivation determines the number of imprint strings enabled and disabled. This means if the inhabiter really wants to change its interaction with the normal matter computer brain for the better, then the inhabiter must choose the psyche altruistic. The inhabiter's continued beneficial choices are displayed in the cognitive states of the lucid dream of the psyche altruistic, and the inhabiter is rewarded for its efforts with sustained uncontrived happiness that comes about from a deep letting go and assuredness derived from mental stability. With enough non-ego-clinging imprint strings enabled, the inhabiter possesses the ability to formulate a clear framework by which to understand its conditioned existence. Because when the inhabiter is aware of itself and its place in the universe, it understands that what appears in the psyches symbolically represents the deeper dimensions of reality of which the Inhabiter can choose to benefit. To be of benefit, the inhabiter cannot allow itself to cower while interacting with the normal matter computer brain. Rather, with an expanded viewpoint and awareness of its primordial presence, the inhabiter's mission is not to get lost within the matrix of its created cognitive states. Therefore, the inhabiter must train itself and become a highly adept mental detective that deciphers the clues present within the psyches. The clues are within the moving holograms of mental events that reveal the inhabiter's relationship with the normal matter computer brain. While thinking with its created cognitive states and using a relative and unlimited viewpoint, the inhabiter must discern if the equation of interaction is balanced or unbalanced.

Clues

The inhabiter is having a nonlucid, psyche self-centered nightmare when the inhabiter experiences the following:[308]

- constant worry, panic, or fear that is nonspecific or associated with rejection, death, failure, contamination, or harming oneself or others
- the need for predictable order and certainty
- a repetitive focus while trying to figure out why it is depressed and feels so bad
- rumination about past mistakes
- intrusive violent, sexual, and religious images formulating a worst-case scenario
- intrusive impulses and urges
- the thought *I've had that thought; if others knew that I had this thought, they would think I was sick*
- frustration while attempting to suppress or neutralize cognitive states
- an increase in the frequency of distressing thoughts, memories, and physical sensations.
- the volume of thoughts flowing through cognitive states are turned up, and the inhabiter becomes cognitively self-conscious with heightened awareness of thoughts
- disturbing thoughts as being fundamentally important
- an overimportance of thought belief, meaning the inhabiter appraises a thought as meaning something more than just a thought (i.e., it thinks that the thought says something about itself as the pawn; because of this misapprehension, the inhabiter identifies with the thought and feels that the thought needs to be dealt with)
- disgust or depression because of its thoughts, which indicates that the inhabiter is apprehending them as being somehow meaningful and important
- *the intrusive-thought theme*: the inhabiter's belief that if it thinks long and hard enough, then the feared outcome will be prevented

(thus the inhabiter engages in mental neutralizing behavior and will feel the upset that comes from trying to neutralize an intrusive thought away; concurrently, the inhabiter might perceive neutralizing behaviors exhibited by the pawn, which includes doing a feared action or fleeing from a situation)

- relief when the inhabiter perceives the pawn drinking alcohol to excess or using illegal drugs
- exhaustion, but the inhabiter feels driven and perceives the pawn engaging in endless activity
- catastrophic thinking related to social situations and a reoccurring question: "Does this person like me?" or "I bet this person doesn't like me" or "Will this person like me?"

The inhabiter is having a nonlucid, psyche self-centered dream when the following occurs:

- The inhabiter believes itself to be the pawn.
- The inhabiter's primary motivation is to make sure the pawn's needs and desires are met, and this is the most important priority within its cognitive states.
- The inhabiter perceives the pawn's actions causing the death of other living beings (human/nonhuman); however, the inhabiter feels indifference to their suffering.
- The inhabiter perceives the pawn's loving words and compassionate actions being extended only to the pawn's friends and family.
- The inhabiter experiences anxiety when the pawn's status, wealth, and possessions are threatened.
- The inhabiter, thinking of the pawn as itself, hopes for its happiness, gain, praise, and fame.
- The inhabiter, thinking of the pawn as itself, fears suffering, loss, blame, and insignificance.
- The inhabiter, thinking of the pawn as itself, feels justified when the character is perceived gossiping or engaging in slander and violence.
- The inhabiter, thinking of the pawn as itself, experiences pride when the character's actions are praised by other humans, and

the pawn's very existence is important because it is connected to those of status.

- The inhabiter, thinking of the pawn as itself, knows its words to be dishonest but classifies them as being necessary *spin*.
- The inhabiter believes its life is over when the pawn dies.

The inhabiter is having a lucid, psyche altruistic dream when the following occurs:

- The inhabiter knows that its existence exceeds that of individuality.
- The inhabiter is vividly aware of its primordial presence.
- The inhabiter does not identify itself as being the pawn.
- The inhabiter understands the complexity and implications of a nonseparable universe and is determined to be of benefit.
- The inhabiter understands that it has primordial consciousness that is always present but cannot be perceived via fleeting sense impressions and mental events.
- The inhabiter accepts its past mistakes when it chose the psyche self-centered, but instead of being a source of depression, the inhabiter is now motivated by what it has learned and uses this to deepen its odyssey of discovery while interacting with normal matter.
- The inhabiter realizes that it has little control over which thought or physical sensations emerge within its cognitive states. Therefore, the inhabiter appraises the problematic thoughts as being fundamentally unimportant and watches them rise and pass away in soft focus.
- The inhabiter has identified a predictable normal matter computer brain computation: if the inhabiter tries to alter the computer brain's circuitry abruptly or tries to block normal matter from performing its processing, a metaphorical bomb will be released in the form of the neurotransmitter program norepinephrine. In abundance, norepinephrine not being hardwired to remain in one place will move in and out of the brain's circuitry. When enabled imprint strings then receive signals from the computer brain and transform the neurotransmitter data and incorporate them into

electromagnetic waves, the inhabiter will experience a feeling that precedes fear and panic. Thus the inhabiter wisely uses its training and remains calm while witnessing cognitive states.

- The inhabiter perceives the pawn regularly engaged in meditation and contemplation while rarely if ever involved in mindless chatter.
- The inhabiter is comfortable with uncertainty.

The inhabiter is having a lucid, psyche altruistic therapeutic nightmare when the following occurs:

- The inhabiter learns from its past mistakes.
- The inhabiter experiences its past choices of the psyche self-centered as a slip. Rather than catastrophizing this previous selection, the inhabiter uses it as a motivation to get back on track.
- The inhabiter is determined to address the cognitive problematic perhaps with a psyche altruistic, lucid therapeutic nightmare. The inhabiter musters the courage to do this by simply reminding itself that repetitive exposure to upsetting or intrusive thoughts (obsessions) will eventually lead to a reduction in discomfort (habituation) and extinction of the obsession.
- The inhabiter stops making faulty assumptions about its feelings and the computer brain's processing. Accepting feelings for what they are (cognized transformed neurotransmitter data), the inhabiter trains itself to acknowledge the preceding pleasant/unpleasant feeling as it is, without cognitively connecting it to a sense impression or mental event.
- The inhabiter's response to physical sensations and thoughts, memory, emotions, and impulses/urges is devoid of censorship, judgment, or modification. The inhabiter ceases to obsessively analyze what was previously problematic by recognizing them to be *mental static*, a by-product of its interaction with a normal matter computer brain, and releases them.

Admittedly, as an individual, the inhabiter perceives information the normal matter computer brain generates on a momentary basis. Relatively, the inhabiter is a focal point of dark matter; thusly, the inhabiter is able to adopt an individualistic viewpoint of its interaction with the normal matter computer brain. However, the inhabiter is a small portion of the universe's dark energy substrate layer that has transformed itself into an unknown number of focal points of dark matter. Accordingly, there are no individuals, only dark energy in its noninteractive form as the universe's substrate layer and in its active form as dark matter inhabiters. This allows the inhabiter to adopt an expanded viewpoint of itself interacting, not as a single inhabiter but as the entire population of inhabiters embodied within normal matter computer brains.

The computer brain receives and shares data with the network of computer brains spread throughout the universe. Thus, while the computer brain processes historical data and while generating new information, the clues that emerge symbolically within the psyches and clearly represent an unbalancing of the equation can be interpreted by the inhabiter as an individual or as itself represented by the population of inhabiters within the universe.

Historical clues that represent the unbalancing of the equation of interaction are as follows: a world with a long history of wars and atrocities, inequalities displayed as slavery, cultural obliteration and extinction of animal species where the entity with the least ability to defend itself consistently suffers. Prosperity that has depleted natural resources while simultaneously producing disposable products that do not simply disappear but negatively impact the environment.

As the normal matter computer brain vacillates between generating the autobiographical-I character and core-I character, historical computations and present-moment processing are always connected and affect the means by which the meta-algorithm modulates and links agents. The inhabiter perceives the connection between the computer brain's previous and current processing where the equation of interaction, having been unbalanced, shapes present events.

Present-moment clues that represent the unbalancing of the equation of interaction are as follows: a world with many wars and atrocities; inequities displayed as poverty; extinction of species and intolerance of cultural diversity; living beings consisting of both humans and nonhumans species with the least ability to defend themselves continually suffering; depleted natural resources with an abundance of toxic waste buried in landfills and global warming.

Although all inhabiters represent dark energy in its interactive form, each has its own population of enabled ego-clinging imprint strings to contend with and a normal matter computer brain that it interacts with. This means though inhabiters communicate with one another via the phenomenon of entanglement and share their experiences, the shared information may help or hinder an inhabiter's ability to balance the equation of interaction and thereby may improve or create chaos in its relationship with normal matter.

There is a fundamental rule as it relates to the equation of interaction: the energy with primordial consciousness must use this gift skillfully to create a harmonious interaction before dark matter and normal matter are enduringly freed from their interaction. Only the inhabiter has a focal point of dark energy as part of its configuration and has primordial consciousness yet also has enabled and disabled imprint strings. Imprint strings are a part of the inhabiter and are not annihilated by the actions of the opposite string; to do so would be self-harming to the inhabiter. However, imprint strings keep a tally of the interaction between the inhabiter and the normal matter computer brain when they are enabled and disabled.

No single theory will predict all the incalculable and unique versions of illusory reality observed by inhabiters within psyche self-centered and psyche altruistic. The number of inhabiters exceeds that of existing stars,[309] and each makes choices that are not predetermined but rather impacted by their enabled and disabled imprint strings. What is similar between inhabiters is that their enabled non-ego-clinging imprint strings are manipulating the environment within normal matter computer brains, guiding inhabiters to a deeper understanding of the nonseparable universe.

Key Point

- A mistaken perception of reality in identifying yourself as the pawn may make you feel as if you are in control of your life; however, clinging to this illusion will not be reassuring when the normal matter computer brain computes for impermanence.

The psyches are comprised of sense impressions and mental events and are created via four steps:

signals received (step one), signals transformed (step two), primordial awareness (step three), and feeling tones and response (step four). The inhabiter's cognitive cycle is made up of two rotations of the four steps, and the inhabiter perceives the steps as a continuous flow of cognitive experience.

Gross and subtle consciousness are within both psyches, and enabled imprint strings will create sense impressions before mental events. This means sense impressions created will have completed the four-step sequence before the electromagnetic waves used to conceive them become the building blocks by which enabled imprint strings create mental events.

Step One: Signals Received

The first rotation: Enabled imprint strings will select the data that pertains specifically to the computer brain to create sense impressions of gross consciousness.

The second rotation: Enabled imprint strings will select the data received by normal matter from the network of interconnected computer brains spread throughout the universe to construct mental events present within subtle consciousness.

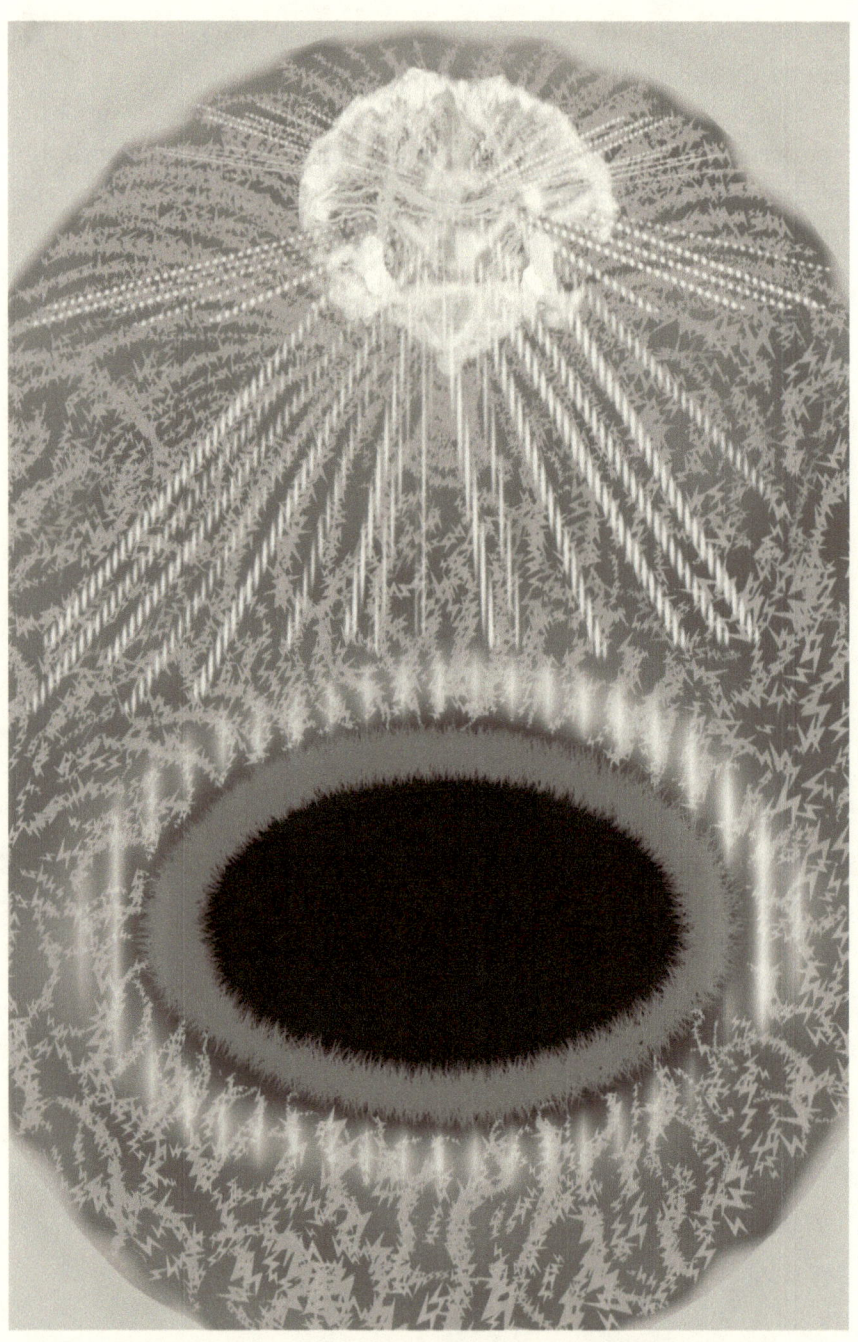

Enabled imprint strings receiving signals and selecting data.

Step Two: Signals Transformed

During the first and second rotation, enabled imprint strings—having received the computer brain's signals—transform them into electromagnetic waves. Therefore, electromagnetic waves contain information that pertains, but is not limited to, neurotransmitter programs involved in the creation of particular computer brain data. Enabled ego-clinging imprint strings oscillate in a way that is directed by the claustra's meta-algorithm, while enabled non-ego-clinging imprint strings oscillate in a way that is similar to enabled ego-clinging imprints, until they tap into the focal point of dark energy. The inhabiter's focal point of dark energy is not directly connected to the meta-algorithm and therefore relies on enabled non-ego-clinging imprint strings to share information they receive from the normal matter computer brain when these strings tap in.

The image displays electromagnetic waves, and the vibrant white depicts transformed neurotransmitter data incorporated into the waves.

Electromagnetic waves with incorporated transformed neurotransmitter data.

Electromagnetic radiation consists of electromagnetic waves containing information that includes neurotransmitter programs involved in the creation of particular computer brain data. At the completion of step two, the electromagnetic radiation of both psyches travels across the surface of the focal point of dark energy and is moved by the kinetic energy produced by oscillating enabled imprint strings. Enabled imprint strings oscillate a specific way when moving the electromagnetic radiation from place to place on the surface of the focal point of dark energy, and this differs depending on whether the first or second rotation.

First rotation: Once the electromagnetic radiation reaches the interior surface of the focal point of dark energy, enabled imprint strings continue to oscillate but temporarily cease to do so in a way that moves electromagnetic radiation or produces kinetic energy. Without kinetic energy produced by enabled imprint strings, the surface of the focal point of dark energy is temporarily smooth. This allows the primordial consciousness of the dark energy focal point to have a clear vantage point by which to analyze the information contained within electromagnetic waves without the rippling effect produced by kinetic energy.

Electromagnetic radiation on the interior surface of the focal point of dark energy.

The electromagnetic radiation of both psyches travels in a similar fashion as it did during the first rotation; however, the oscillations of enabled imprint strings increase, and this causes the electromagnetic radiation of the two psyches to coil upon itself to produce layers and creates what could be conceived as a funnel shape of layered electromagnetic radiation. Enabled imprint strings' continued oscillations move electromagnetic radiation of the two psyches and also cause the funnel-shaped electromagnetic radiation to levitate just above the surface of the focal point of dark energy.

The funnel-shaped electromagnetic radiation has a narrowed base that widens with increasing layers. The base of the layered electromagnetic radiation contains the least amount of electromagnetic waves with incorporated transformed neurotransmitter data and represents the gross conscious level of sense impressions (physical sensations). As the layered electromagnetic radiation gets higher, it widens and contains more electromagnetic waves, with incorporated transformed neurotransmitter data representing the subtle conscious level of mental events (thoughts, memory, emotions).

Because the narrowed base of the funnel-shaped electromagnetic radiation does not directly make contact with the surface of the focal point of dark energy but levitates just above it, a space is created between the focal point's surface and layered electromagnetic radiation. This space represents the extremely subtle conscious level that is devoid of electromagnetic waves of sense impressions and mental events.

1. Extremely subtle conscious level
2. Gross conscious level
3. Subtle conscious level

Levitating layered electromagnetic radiation.

Step three: During the first and second rotation of the four-step sequence, the inhabiter uses its primordial consciousness derived from its focal point of dark energy and is aware of the electromagnetic waves of both the psyche self-centered and the psyche altruistic. This means regardless of how many or what kind of neurotransmitters are released by the computer brain at any given moment, when transformed by enabled imprint strings and incorporated into electromagnetic waves, there is not one processing event that escapes the awareness of the inhabiter's focal point of dark energy. The amount of primordial intelligence that emanates from the focal point of dark energy does not change; whether illuminating the surface of the focal point of dark energy, extremely subtle conscious level or illuminating the gross or subtle conscious levels of the layered electromagnetic radiation, the amount is the same.

During the first rotation, primordial consciousness illuminates the surface of the focal point of dark energy and is depicted in this image.

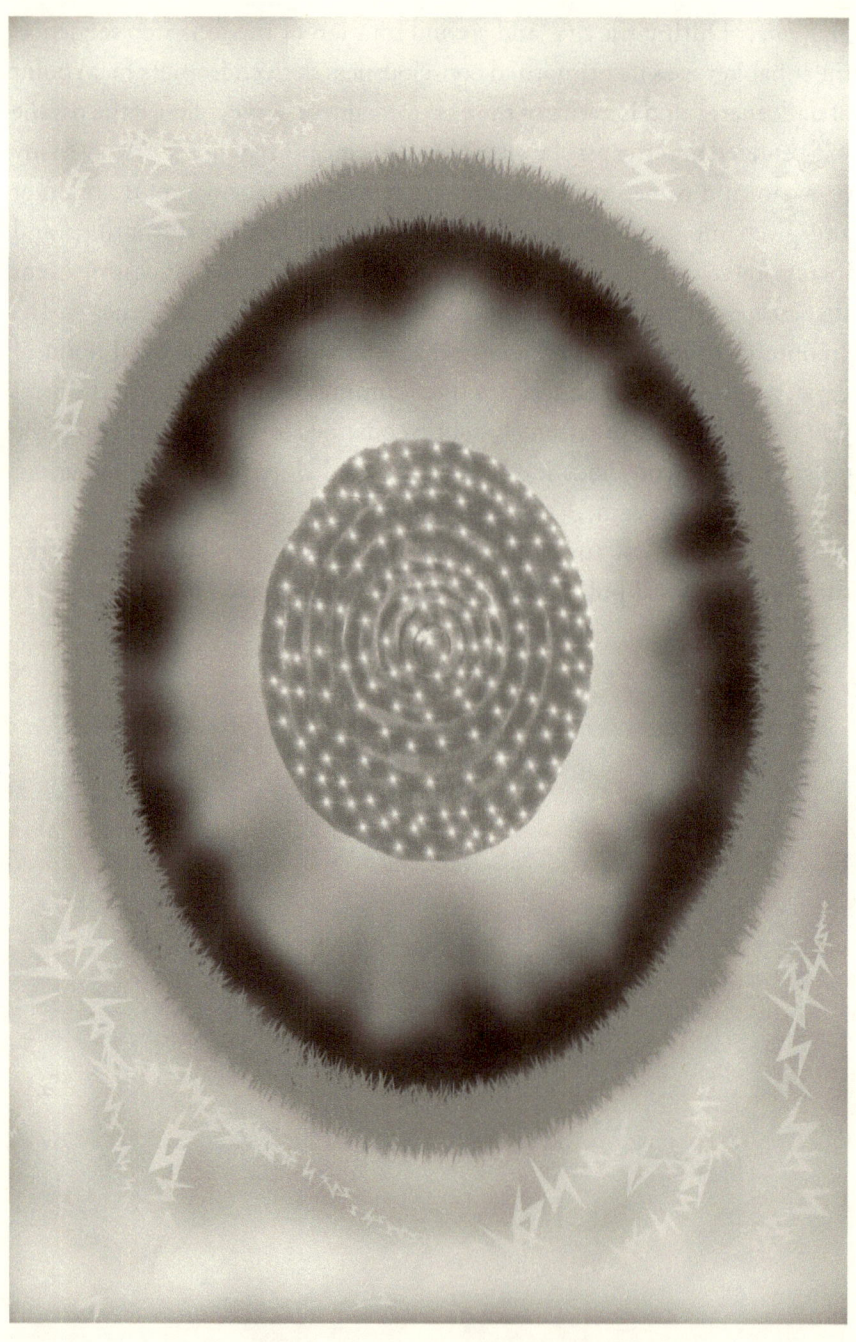

The surface of the focal point of dark energy illuminated
by the inhabiter's primordial consciousness.

In step three during the second rotation, there is kinetic energy on the surface of the focal point, except for a centralized area located beneath the base of the layered electromagnetic radiation where there is no kinetic energy present. Thus, an iris/pupil pattern is created: where kinetic energy is present on the surface of the focal point of dark energy, the emission of primordial consciousness is blocked; however, it flows from the centralized area devoid of kinetic energy. This yields a phenomenon in the form of a beam of primordial consciousness. The beam arises from the focal point and travels in an upward direction, passing through the layers of the electromagnetic radiation to the subtle conscious level. The inhabiter's beam of primordial consciousness is more narrowed at the surface of the focal point and spreads out in multiple directions at the subtle conscious level. Because of this, primordial consciousness is able to accommodate the increase in electromagnetic waves at this level.

When transformed neurotransmitter data is detected by primordial consciousness with its beam, it reacts by illuminating electromagnetic waves of both psyches simultaneously, which includes the gross and subtle levels within the electromagnetic radiation and is depicted in this picture.

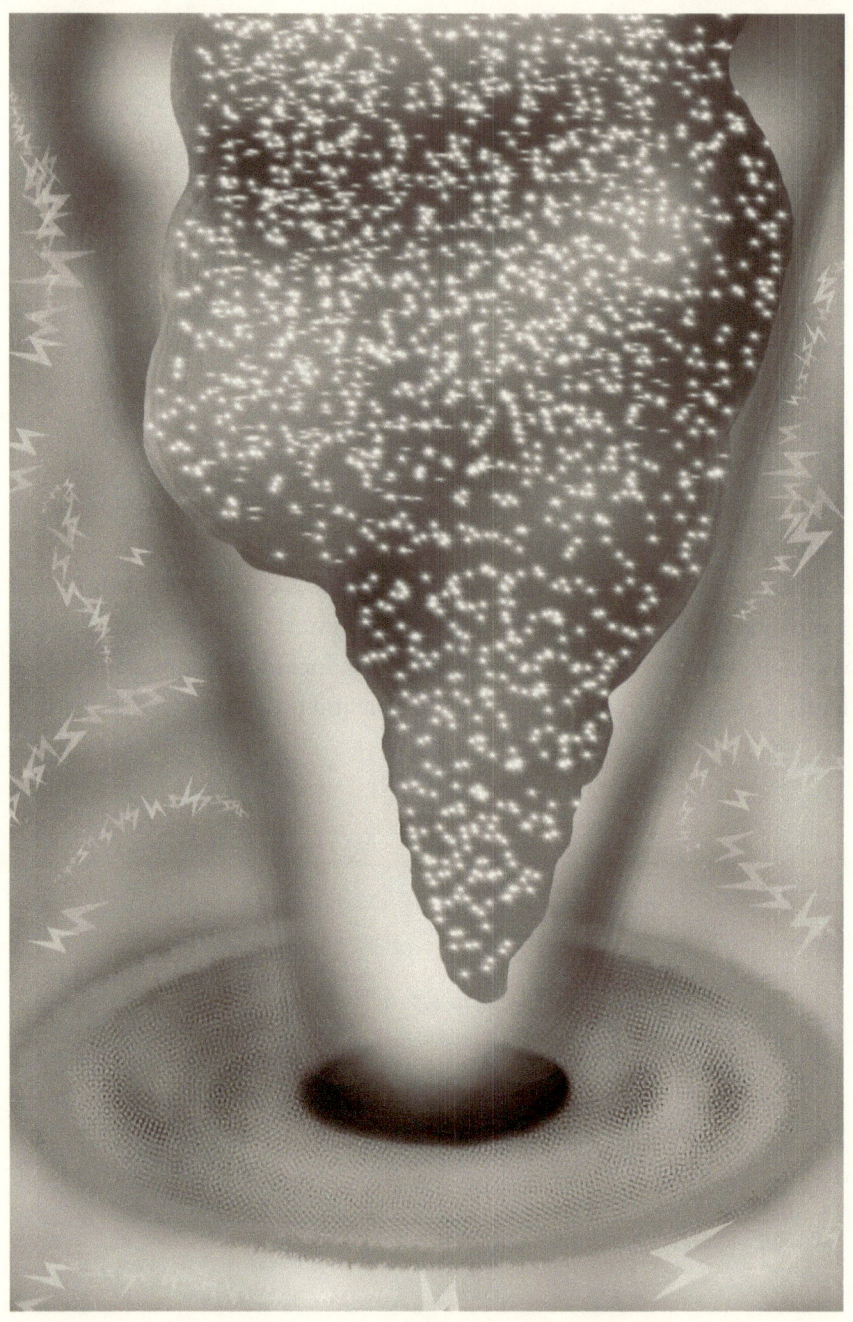

Transformed neurotransmitter data is detected by primordial consciousness.

The inhabiter's beam of primordial consciousness has within it a central region. This region represents a zone of intelligence with the capacity to move up and down to different levels within the layered electromagnetic radiation and in many different directions but in a way that differs from the beam itself.

The vividness of primordial consciousness does not change in the central region; rather, it is coherent intelligence that functions like a laser of primordial consciousness. The movement of the central region of the beam is governed by primordial consciousness's detection of transformed neurotransmitter data incorporated into electromagnetic waves. When transformed neurotransmitter data is detected by primordial consciousness with the central region of its beam, this intelligence will follow where associated processing leads by illuminating the electromagnetic waves created by the most enabled imprint string type during one cognitive cycle (two rotations of the four steps), before repeating this process for the other type of enabled imprint string.

Coherent intelligence that functions like a laser of primordial consciousness is depicted in this picture by the gray lines.

The coherent intelligence of primordial consciousness detecting transformed
neurotransmitter data incorporated into electromagnetic waves.

Step Four: Feeling Tones and Response

Both the first and second rotations are distinct in that the inhabiter thinks differently.

However, unlike when the inhabiter detects transformed neurotransmitter data with its beam and central region of primordial consciousness, while thinking with created conscious states, the inhabiter first becomes aware of transformed neurotransmitter data as the feeling tones of pleasant, unpleasant, or neutral.

First rotation: The inhabiter, in not thinking with a fully developed cognitive state, responds to the sense impressions of gross consciousness by predominately reacting once aware of transformed neurotransmitter data as the feeling tones of pleasant, unpleasant, or neutral, versus cognizing its choices. A reverberation is produced that begins on the interior surface of the focal point of dark energy and travels to the peripheral surface. When the force of the reverberation reaches the peripheral surface of the dark energy focal point depicted in this picture, this signals enabled imprint strings to use electromagnetic waves of sense impressions as building blocks to create mental events.

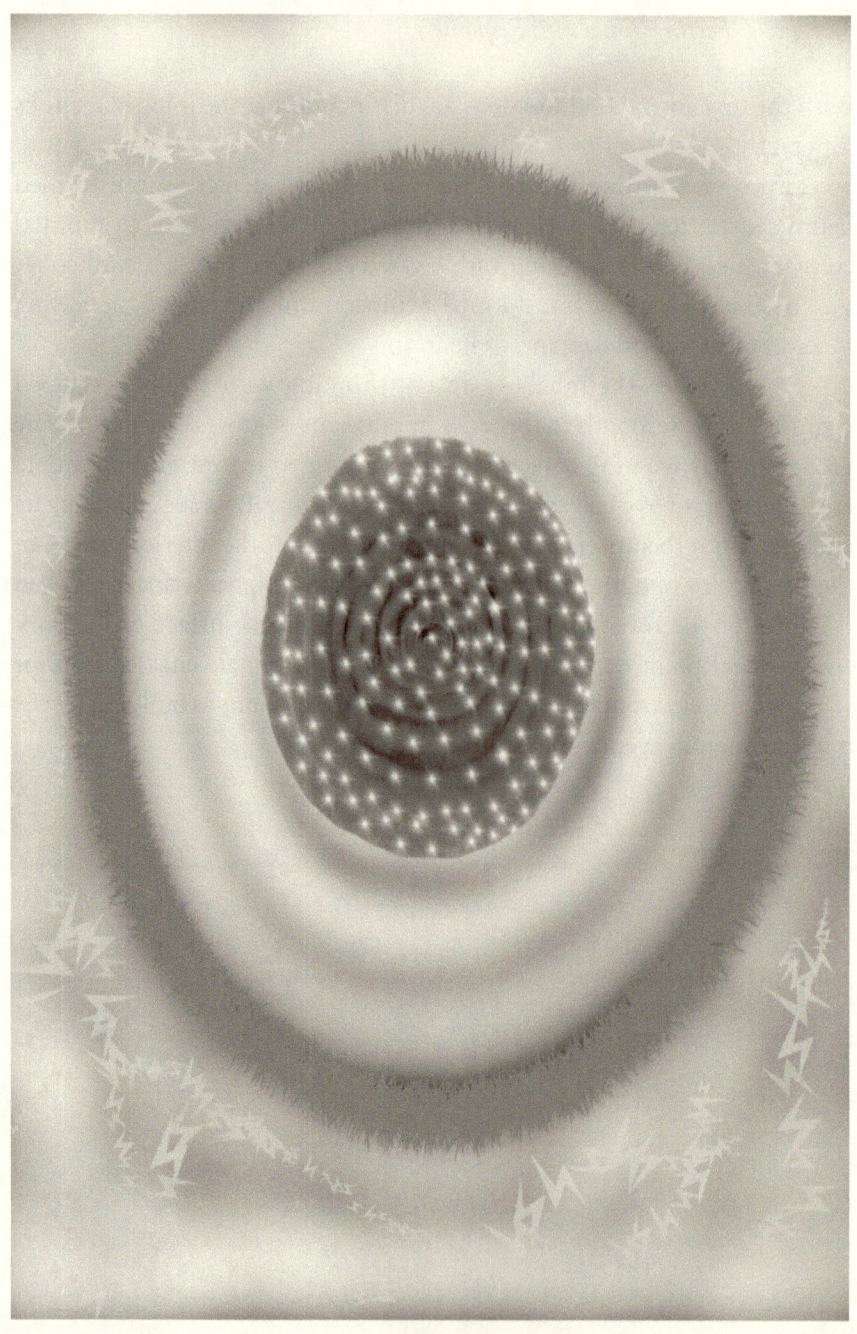

A reverberation.

Second rotation: The inhabiter experience emotions, thinks with thoughts, and has memories consisting of three-dimensional images or moving holograms. The inhabiter perceives electromagnetic waves as mental events that begin in step four, during the second rotation, with awareness of transformed neurotransmitter data as feeling tones pleasant, unpleasant, or neutral.

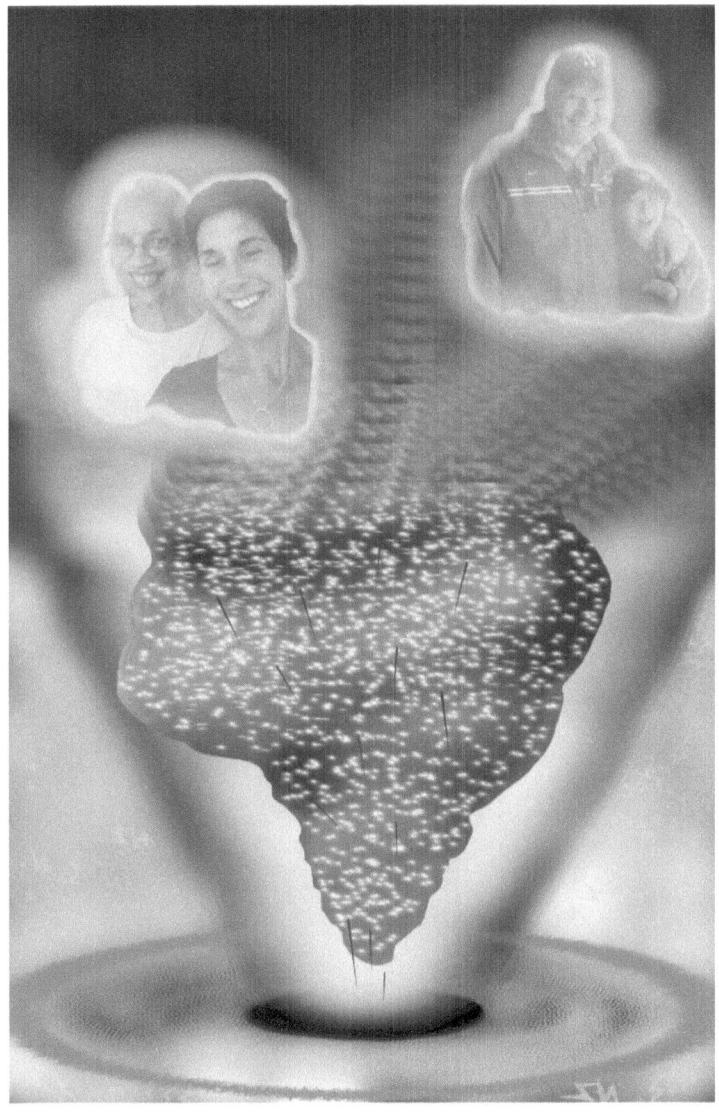

The inhabiter thinks.

The inhabiter selects one of the two psyches to experience its momentary reality, and in doing so, the inhabiter thereby is responding to the psyche in a way that disables and enables its imprint strings. The electromagnetic radiation of both psyches dissolves into the focal point of dark energy.

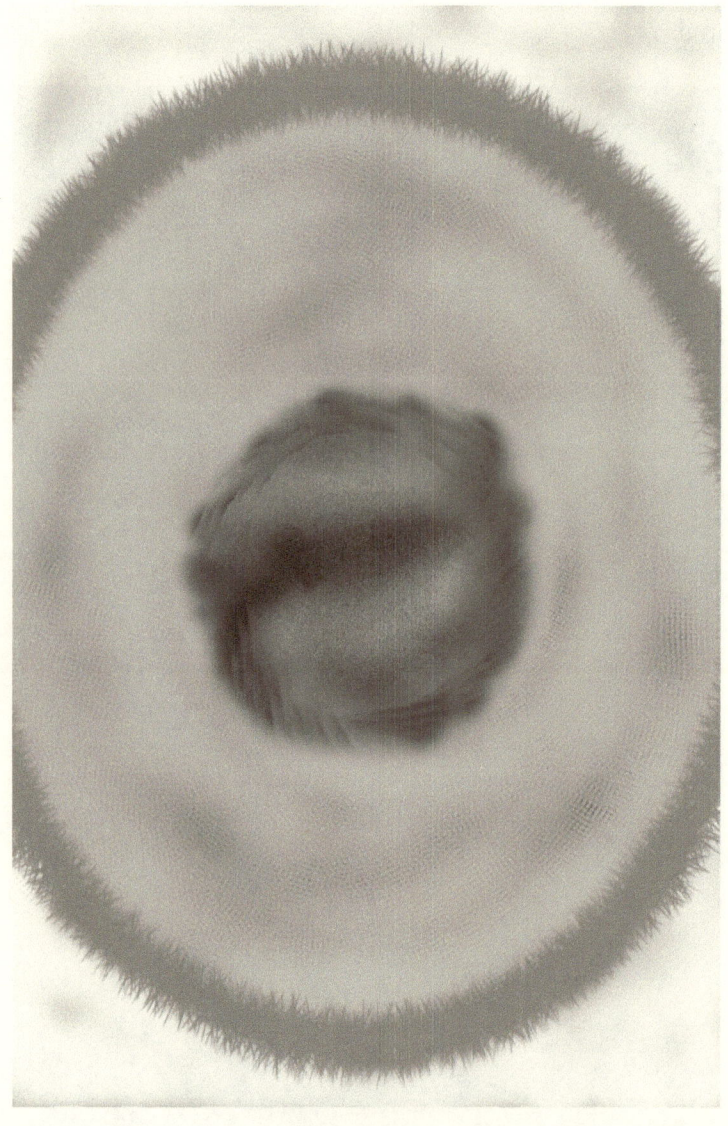

Looking down on the focal point of dark energy where the electromagnetic radiation is in the final stages of dissolving into the focal point.

The electromagnetic radiation of both psyches having dissolved into the focal point of dark energy causes a reverberation with a force that correlates directly with the inhabiter's motivation. Subsequently, the reverberation moves across the surface of inhabiter's focal point of dark energy until it reaches the enabled imprint strings and will determine the type of imprint strings and how many of them will be disabled. The enabled imprint strings that are left will transmit alternating signals of magnetic charges that will be picked up by the normal matter computer brain's antenna of the brain stem and spinal cord and will be received by the meta-algorithm in the interclaustral pathway.

When imprint strings are disabled, a single photon of light is produced, and both types of imprint strings, when enabled, have opposite charges. As photons are produced by the disabling of imprint strings, they will affect the forces within the electromagnetic radiation and correspondingly the inhabiter's perception.

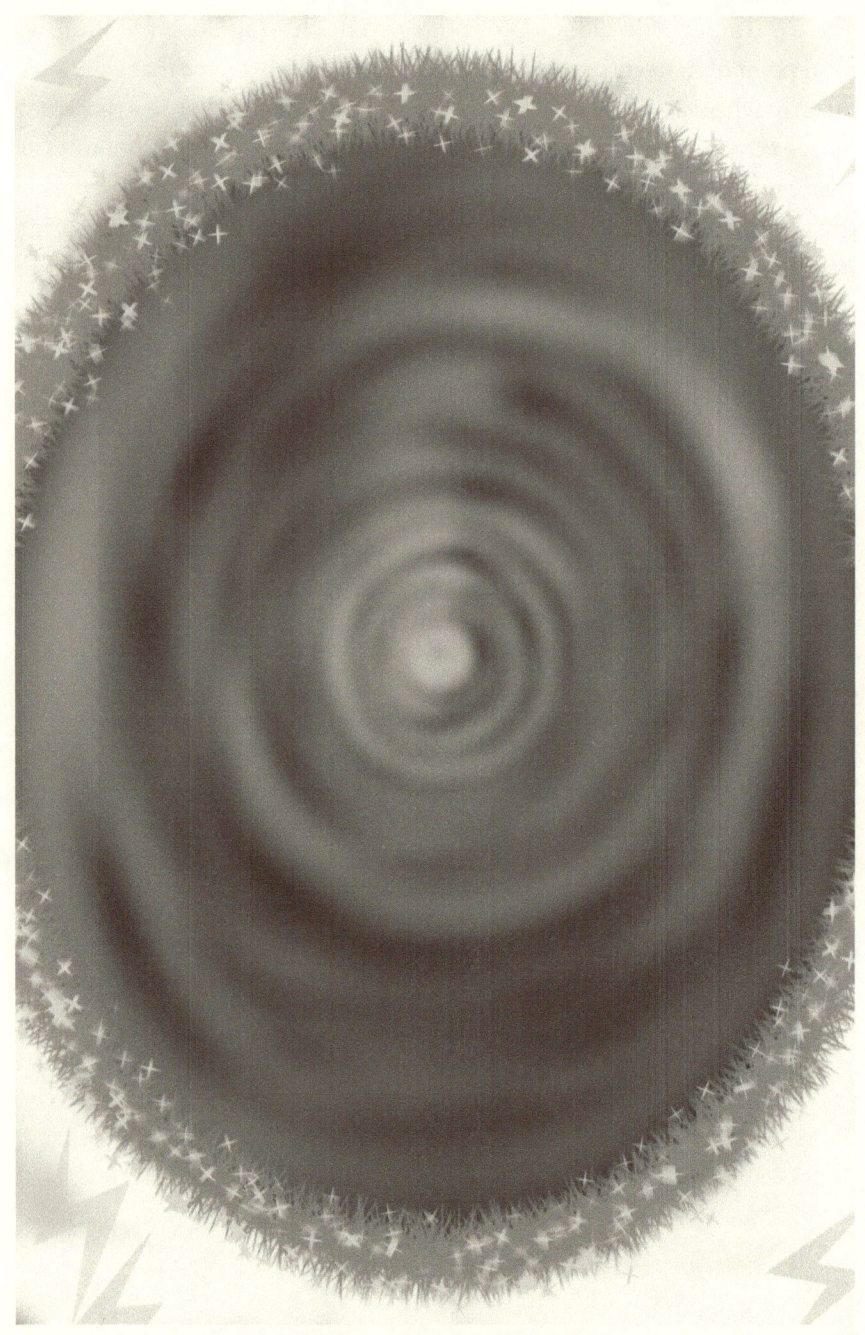

Reverberation and photons produced.

The universe's dark energy substrate layer acts as a two-dimensional grid, and within that grid are an unknown number of attached focal points of dark energy. The focal point of dark energy, having absorbed the electromagnetic radiation of both psyches during step four of the second rotation, transforms it to a discrete quantity of energy as a signal to the universe's dark energy substrate layer that is proportional in magnitude to the frequency of the radiation it represents. The universe's substrate layer of dark energy relays signals of quantum information among inhabiters via the phenomenon of entanglement. The signals sent by the universe's dark energy substrate layer contain information about the inhabiter's I character, and each inhabiter communicating will be receiving information at the same time, which creates the illusion in the psyche to the inhabiter observing it that its I character is interacting with other living beings.

Entanglement.

The relationship between the inhabiter and the computer brain is special because of the way these two energy forms interact and the fact that the inhabiter is the only observer of space and time. The concept of time is the reference by which the inhabiter contextualizes its mental experience. Time commences when the primordial beam of awareness emanates from the surface of the focal point and reaches the subtle conscious level and when the central region of the beam illuminates the electromagnetic waves of a psyche.

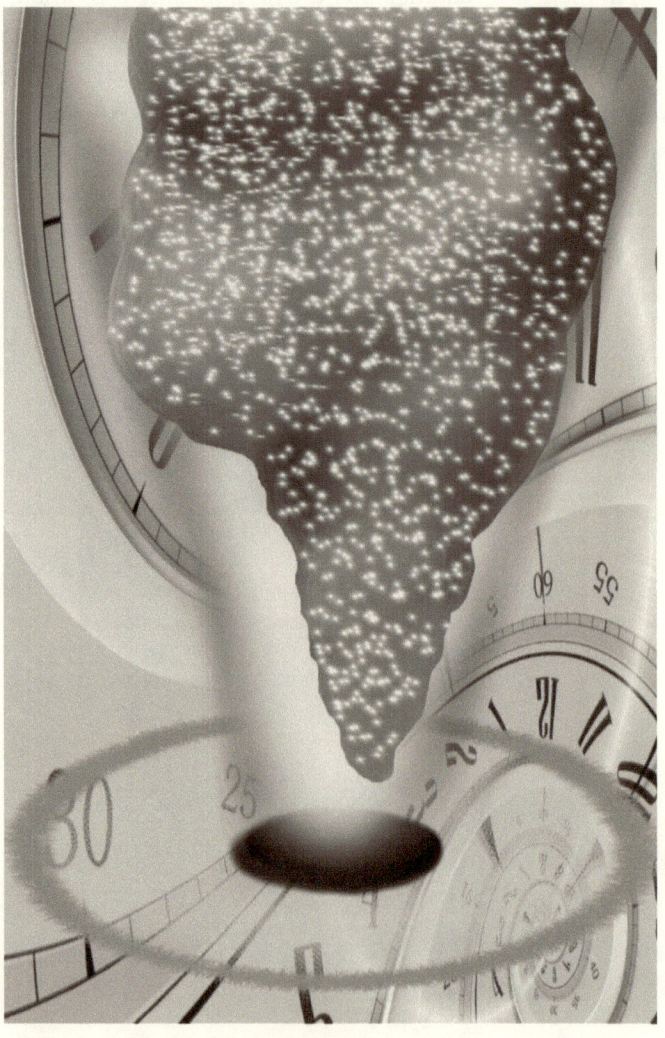

Time.

Reminders before Reading Chapter 11

Neurotransmitters/Transformed Neurotransmitter Data and Related Topics

- neurotransmitters: The computer brain's communicative programs and the means by which neurons are able to transfer information as a nerve impulse to another nerve fiber, a normal matter muscle fiber, and other normal matter structures.
- transformed neurotransmitter data: Neurotransmitter data that are a part of the computer brain's signals received and transformed by enabled imprint strings. Transformed neurotransmitter data acts as a signal within electromagnetic waves that draws the attention of the inhabiter's primordial consciousness and determines its reaction.
- feeling tones: Transformed neurotransmitter data that will be experienced by the inhabiter as the feeling of pleasant, unpleasant, or neutral that are associated with physical sensations and mental events when the inhabiter is thinking with cognitive states.
- mental bond: A tether created by transformed neurotransmitter data that serves to connect the inhabiter's primordial reactions to the inhabiter's cognitive reactions while thinking with created cognitive states.

Primordial Consciousness and Related Topics

- primordial consciousness: The inhabiter's intelligence derived from its focal point of dark energy that is nonconceptual awareness capable of analyzing information such as forces and transformed neurotransmitter data contained within electromagnetic waves. The inhabiter's primordial consciousness reacts by illuminating electromagnetic waves with intelligence that allows the inhabiter to think with created cognitive states.

- focal point of dark energy: A part of the inhabiter's configuration that possesses primordial consciousness. This intelligence illuminates in all directions within the inhabiter's focal point of dark energy but also emanates from the surface of the focal point.
- iris/pupil pattern: A pattern on the surface of the inhabiter's focal point of dark energy that is produced by the increased oscillations of enabled imprint strings that create kinetic energy, except for a centralized area located beneath the base of the layered electromagnetic radiation.
- beam of primordial awareness: A phenomenon that occurs when the inhabiter's beam of primordial consciousness is blocked in some directions by the kinetic energy produced on its focal point of dark energy. However, the beam of primordial consciousness flows freely from the centralized area just beneath the levitating electromagnetic radiation that is devoid of kinetic energy. The beam arises from the surface of the focal point of dark energy and travels in an upward direction, passing through the layers of the electromagnetic radiation to the subtle conscious level, illuminating the electromagnetic waves of both psyches simultaneously.
- central region: A zone within the beam of the inhabiter's primordial consciousness with the capacity to move up and down to different levels within the layered electromagnetic radiation in many different directions but in a way that differs from the beam itself.

Electromagnetic Radiation and Related Topics

- layered electromagnetic radiation: Produced by the increase in oscillations of enabled imprint strings and consists of two psyches coiled to form layers and a funnel shape of electromagnetic radiation.
- funnel-shaped electromagnetic radiation: Levitates just above the surface of the inhabiter's focal point of dark energy.

- levels of the layered electromagnetic radiation:
    - Extremely subtle conscious level is the space created between the surface of the inhabiter's focal point of dark energy and the base of the funnel-shaped electromagnetic radiation that is devoid of electromagnetic waves.
    - Gross conscious level is located at the base of the layered electromagnetic radiation that contains electromagnetic waves of sense impressions.
    - Subtle conscious level is the wider and higher portion of the layered electromagnetic radiation that contains electromagnetic waves of mental events.

Psyches

- psyche self-centered: When the inhabiter chooses this psyche, then the inhabiter is having a nonlucid dream or a nonlucid nightmare.
- psyche altruistic: When the inhabiter chooses this psyche, then the inhabiter is having a lucid dream or a lucid therapeutic nightmare.

Miscellaneous Topics

- thinking: The inhabiter must use either the cognitive states within psyche self-centered or psyche altruistic, not both simultaneously, to think. The inhabiter will perceive all four steps used to create the psyches via two rotations of the steps as a continuous flow of cognitive experience.
- time: A reference by which the inhabiter contextualizes its mental experience and emerges only within the psyches. Time commences when the inhabiter's primordial beam of primordial consciousness emanates from the surface of the focal point and reaches the subtle conscious level and when the central region of the beam illuminates the electromagnetic waves of a psyche.

- What is real? When the inhabiter chooses one of the psyches to experience its momentary reality, the cognitive states of that psyche define situations as real.
- the game of life: A term used to describe the inhabiter's life of conditioned existence while embodied within a normal matter computer brain. The inhabiter participates in the game of life by choosing a psyche by which to experience momentary reality.
- pawn: The I character that is the avatar of interaction between the inhabiter and the normal matter computer brain.
- motivation: Determines the number of imprint strings enabled and disabled.
- clues: Exist within the moving holograms of mental events at the subtle conscious level of the layered electromagnetic radiation. When the inhabiter assembles clues, then the inhabiter has the ability to discern which psyche it is perceiving while thinking with the cognitive states of that psyche.

CHAPTER 11

The Game of Life

Have you ever had a dream you thought was real?[310] Profoundly, this is the viewpoint shared by many inhabiters, though they are unaware of it, while participating in many rounds of the game of life, lost in a nonlucid dream or nightmare of the psyche self-centered. Inhabiters, having chosen the psyche self-centered while embodied within a normal matter computer brain, regularly unbalance the equation of interaction. Every inhabiter at some point in its journey interacting with normal matter has been—and many continue to be—immersed in the cognitive states of the psyche self-centered. There are many mysteries surrounding the universe, and only bits and pieces are well-known.[311] What is not an enigma is that there are variables built into the equation of interaction that address extreme cases where an inhabiter, through its choices, has unbalanced the equation of interaction:

Scenario one. If an inhabiter chooses the psyche self-centered but does not have the quantity of its ego-clinging imprint strings to fulfill the demand of enabling them, the inhabiter's focal point of dark energy will simply morph and create more of them. This example represents a dynamic variable built into the equation of interaction, and the inhabiter's imprint string population can increase in number as needed to fulfill a demand.

Scenario two. If an inhabiter has managed to disable so many of its non-ego-clinging imprint strings that there is only one of this type of imprint string enabled, this would equate to a monumental unbalancing of the equation of interaction. Nevertheless, dark energy did not construct the variables in the equation of interaction to remain in conditioned existence. Thus dark energy, having anticipated that as a dark matter inhabiter it would become immersed while interacting with normal matter, earmarked as a safety measure a single non-ego-clinging imprint string. This enabled non-ego-clinging imprint string prevents any dark matter inhabiter from maximally unbalancing the equation of interaction to such an extreme that normal matter would be unkindly driven to an eternal engagement. If an inhabiter has only a single enabled non-ego-clinging imprint string, its focal point of dark energy prevents this imprint string from being disabled until the inhabiter manages to enable more of these imprint strings. Although dark energy accepted the risks entailed in the game, it did not enter into a life of conditioned existence as a dark matter inhabiter foolishly. This is why the universe's dark energy substrate layer morphed to create two types of imprint strings. The unsurpassed intelligence of the universe's dark energy ultimately controls its life experience, and the inhabiter's configuration did not manifest by chance; costs were weighed, and a focal point of dark energy and non-ego-clinging imprint strings were included in each inhabiter's configuration. These design details act as variables within the equation of interaction that prevents the dark matter inhabiter from becoming completely immersed with normal matter without the means by which to enduringly free these energies from their interaction. After the inhabiter has gathered enough pertinent information through experience, enabled non-ego-clinging imprint strings reverse the inhabiter's immersive experience. The inhabiter's primordial reactions and cognitive reactions are unified while the inhabiter consistently responds by selecting the psyche altruistic. While thinking, the dark matter inhabiter uses its primordial consciousness to guide the process of reemergence, and this is depicted in the cognitive states of the psyche altruistic. Reemergence is a process though and does not happen in an instant; instead, it is created through many transformative imperceptible waves of cognition that affect both the dark matter inhabiter and the normal matter computer brain because of how these two energies are interacting.

While inhabiters are embodied within normal matter computer brains and participate in the game of life, they will think uniquely with created cognitive states. What allows an inhabiter to think with cognitive states is the fact that its configuration includes a focal point of dark energy with primordial consciousness and enabled imprint strings. In this way, each new round of the game of life could be considered an inhabiter's continua of thinking that could be compared to waves in the ocean. Imagine that enabled imprint strings are particles in the ocean that produce a gravitational force that pulls normal energy to the inhabiter's focal point of dark energy, thereby producing waves in the water. When normal energy reaches the inhabiter's focal point of dark energy, it is transformed to normal matter by moving in circles that generate swells; however, the particles (enabled imprint stings) in the ocean don't travel to the beach but rather remain oscillating while attached to the inhabiter's focal point of dark energy. The inhabiter's focal point of dark energy with enabled imprint strings might be compared to the inhabiter's ability to transmit knowledge from one round of the game of life to another. This knowledge cannot be called a thing; it instead equates to being more like a consciousness wave associated with each inhabiter. In the same way that a given wave rolling on the beach is distinct from all the others, the inhabiter's participation in the game of life and the conscious states it will use to think are distinct.[312]

Nevertheless, when the universe's dark energy substrate layer made a heroic gesture and embarked on a journey in the form of a dark matter inhabiter, it spawned an entire race of normal matter computer brains.[313] The number of dark matter inhabiters that initially began their odyssey of discovery together, and how the ongoing process of one inhabiter completes its journey and another begins, is unknown. One inhabiter emerging out of the countless many creates an imperceptible amount of change, and the minuscule amount of normal matter reverting to its invisible form as normal energy is undetectable. Albeit, when a focal point of dark energy morphs into dark matter with infinitely small oscillating energy[314] that are imprint strings, an inhabiter participates in the game of life that begins with an event that might be considered birth into the bondage[315] of conditioned existence. The inhabiter is a visitor in the vessel of the normal matter computer brain.[316] When the inhabiter is inhabiting a normal matter computer brain configured to produce data for a human psyche, agents are sequenced by the meta-algorithm, yet it is the inhabiter that sees, hears, smells, tastes, or feels them.[317] At the subtle conscious level of the layered electromagnetic radiation, appearances are deceiving,[318] as there is no psyche that depicts phenomena as electrical patterns of neural synchrony; interacting energies are made to appear solid, and where there is unpredictability, predictable order appears as an illusory universe containing a world that is perceived by a pawn. Thus the inhabiter participates in the game of life that consists of four periods: formation, continuity, destruction, and nonmanifest state.[319]

- Formation: The period where dark energy transforms itself to dark matter, and this causes the transformation of normal energy into normal matter that will embody an inhabiter as a computer brain. Accordingly, formation marks the beginning of an initial interaction between a dark matter inhabiter and a normal matter computer brain. However, formation might also represent an inhabiter's continued interaction with normal matter and the start of another round in the game of life, if the inhabiter was unable to disable its ego-clinging imprint strings.

Many inhabiters have participated in an abundant number of games of life in countless diversely configured normal matter computer brains. However, during each round of the game of life, the inhabiter's enabled imprint strings and primordial consciousness function similarly. Enabled imprint strings create electromagnetic waves that are illuminated by the inhabiter's primordial consciousness. Because enabled imprint strings can only be disabled, not destroyed, and dark energy as a focal point does not have a life span, together these variables of an inhabiter's configuration conspire to produce similar crystallization in the inhabiter's thinking. This comes into play when the inhabiter is embodied by a normal matter computer brain that is similar to a previous configuration. The inhabiter's enabled imprint strings create electromagnetic waves as they did previously; however, not with the exact same data, and the inhabiter's primordial consciousness will illuminate these electromagnetic waves. Accordingly, the inhabiter will more or less perceive its life of conditioned existence in a similar previous fashion, and in this way, some concepts become fixed.[320] Nevertheless, the inhabiter's population of imprint strings that are enabled and disabled are not fixed. Instead, they are in a constant state of flux from one round of the game of life to another and represent dynamic variables in the equation of interaction. When enabled imprint strings oscillate, they communicate a gravitational force to the normal energy close by. This gravitational force is so strong that any normal energy near is pulled toward the inhabiter's focal point of dark energy, and when pulled near, normal energy folds and forms as normal matter around the inhabiter. How normal matter folds around the inhabiter creates the shape and structure of the normal matter computer brain. Thereby, evolution in the universe is actually defined by enabled imprint strings, where a higher-functioning computer brain directly correlates with the data that is produced and transformed by enabled imprint strings to create the electromagnetic waves of a psyche. The more enabled ego-clinging imprint strings an inhabiter has, the less evolved the normal matter computer brain will be that embodies the inhabiter. A human brain has many warps and curves, reflecting its many capacities, where in comparison, a mouse brain has very few.[321] When forming a computer brain with the capacities to compute for a human psyche, normal matter curls upon itself like a tent pulled by ropes with a pole in the middle, and normal matter is formed to connect all parts of what is to become the normal matter computer brain.[322] Collectively, the transformation of normal energy to normal matter forms the human computer brain, a unique and powerful computer made of normal matter that is situated on the universe's dark energy substrate layer. Thereby, inhabiters that are embodied by this sort of configured normal matter computer brain will perceive the human version of the psyche self-centered and the psyche altruistic while thinking with cognitive states that include eight distinct types of intelligence to varying degrees. Thereby, some inhabiters will have the ability to describe the process of how neurons collect information and communicate with one another via mathematics.[323] By doing so, theses inhabiters formulate a relative viewpoint that allows them to understand the fabric of the computer brain's circuitry. It is the normal matter computer brain's design secret of diversity that is revealed through mathematics.[324] Each inhabiter perceives reality differently, in part because of the uniqueness of the normal matter computer brain it inhabits. Within an individual normal matter computer brain, no neuron is the same as another, and when comparing the many different kinds of computer brains present in the universe, there are different sizes. This equates to computer brains differing in the number of neurons they have and the neurons of a particular computer brain not being orientated in exactly the same way as they are in another.[325] However, normal matter computer brains, regardless of size or species, do not see pictures of objects but rather degrees of neural synchrony expressed as electrical activity.[326] Though neurons differ between normal matter computer brains, what is similar is how neurons produce and process information, and there are mathematical formulas that can simulate the activity of a human neocortex, thus giving the inhabiter a glimpse into the life of normal matter.[327]

However, if an inhabiter ended a round of the game of life with a small number of non-ego-clinging imprint strings enabled, its next round interacting with normal matter could be appropriately entitled "hell on earth," especially if the pawn was that of a food animal, endangered wildlife species, designated destructive animal, or insect. This would be because the inhabiter would be limited to thinking with the psyches of these species. Correspondingly, the inhabiter would have limits in its ability to fully comprehend the complexities of the game of life, but the inhabiter would experience many unpleasant physical sensations and mental events.

- Continuity: The period in which the dark matter inhabiter interacts with normal matter while embodied by a computer brain. During this period, the inhabiter balances or unbalances the equation of interaction via its choices. The ability that an inhabiter has to balance the equation of interaction through its choices and disable its ego-clinging imprint strings directly correlates with the capacity of the normal matter computer brain to generate and process data. By virtue of the fact, the psyches are created by enabled imprint strings transforming the computer brain data into electromagnetic waves of sense impressions and mental events. The less complexly configured a normal matter computer brain, the decreased ability of the inhabiter to cognize its choices and the greater consistency that the inhabiter unbalances the equation of interaction. In this scenario, the inhabiter would habitually enable its ego-clinging imprint strings that would increase the inhabiter's suffering, because, with limited means by which to make alternative choices, the inhabiter would experience the psyche self-centered as a nonlucid nightmare. Only when the inhabiter managed to enable its non-ego-clinging imprint strings, perhaps by chance, would its predicament be altered, with the most favorable scenario being when normal matter was configured to produce data for a human psyche. Thus the inhabiter would have the ability to make the alternative choice of the psyche altruistic.

Incorporated within the electromagnetic waves of the human version of the psyche altruistic is the transformed neurotransmitter data that was further altered when enabled non-ego-clinging imprint strings tapped into the inhabiter's focal point of dark energy. When transformed neurotransmitter data is detected by the inhabiter's beam of primordial consciousness within electromagnetic waves of this psyche, the *mental bond* is produced that tethers the inhabiter's primordial reactions to the inhabiter's reactions while thinking with the cognitive states of the psyche altruistic, resulting in mental stability. Thus the inhabiter's primordial reactions are purposeful and alter familiar experiences at the subtle conscious level with contextual factors that gain the inhabiter's attention while thinking with the cognitive states of the psyche altruistic. However, during the process of reemergence, the inhabiter's perception of the pawn as being itself remains consistent, with little alteration in this belief until the inhabiter is able to build up its population of enabled non-ego-clinging imprint strings. Accordingly, the inhabiter gains cognitive strength through the positive force of these magnetic monopoles. With each non-ego-clinging imprint string that stays enabled, the computer brain's programming gets gradually changed in a way that avoids alerting the meta-algorithm but produces consistent alterations to the computer brain's programming. This is the activity aspect of the inhabiter's primordial consciousness that is neither created by neurons nor guided in its actions by the amygdala, thalamus, hypothalamus, or brain stem when illuminating the electromagnetic waves of the psyche altruistic.[328] However, the inhabiter's focal point of dark energy monitors these key brain structures when enabled non-ego-clinging imprint strings tap in and share the information they received as signals from the normal matter computer brain. In this way, activity-dependent plasticity plays an important role in facilitating the inhabiter's transformation into a skilled participant in the game of life.[329] It is a term used to describe synaptic strengthening by correlated activity and sprouting of new connections.[330] The opposite phenomenon is true, and synapses that experience minimal activity will eventually lose their connection and deteriorate.[331] Electrical patterns that pass through synapses contain information representing sensory and motor mapping within the computer brain's implicit layer of circuitry, and it is not entirely hardwired with fixed neuronal circuits.[332] There are many instances of cortical and subcortical rewiring as synaptic connections are removed or recreated, depending on the activity of the neurons that use them.[333]

As the inhabiter uses its primordial consciousness to guide itself from the abyss of its inaccurate perceptions and thinks with the cognitive states of the psyche altruistic, the inhabiter makes progress in reemergence through discovery learning.[334] This type of learning takes place when the central region of the inhabiter's beam of primordial consciousness specifically illuminates electromagnetic waves in a way that recreate scenes within the psyche altruistic that were problematic while the inhabiter was immersed and habitually choosing the psyche self-centered. Thereby, within the cognitive states of the psyche altruistic, there are no mysterious coincidences, and déjà vu moments are simply the display of the inhabiter's primordial consciousness in creative action, which utilizes transformed neurotransmitter data generated by the computer brain's circuitry that is located in hybrid architecture of interconnected networks.

- Destruction: The period that starts when normal matter begins its computation for impermanence and ends when the inhabiter is no longer embodied by a computer brain. As the computer brain configured to produce data for a human psyche starts its computation for impermanence, its temperature changes, and its circuitry begins to malfunction. It is a death spiral of firing neurons and broken connections in which normal matter struggles to function as a unit. Although the computer brain never perceived the pawn like the inhabiter, it will not let this meta-stable construct go willingly but will compute furiously as a way to protect it. The meta-algorithm will do its best to maintain internal states by correlating synchrony of biologic information. As they are not hardwired to remain in one place, there will be a flow of neurotransmitters everywhere and yet nowhere.[335] These communicative programs will move in and out of circuitry, connecting and altering malfunctioning neurons in a variety of ways.

As the many layers of the computer brain's interconnected networks are crashing, the inhabiter will be a witness to its neural meltdown, until the only computation afforded by normal matter will be for the most basic but potent information. The link to information flow between the computer brain's implicit and explicit layer of circuitry will be controlled in part by the thalamus. The thalamus, as the computer brain's train station, will allow programs in and out of the implicit layer of circuitry, including those from the brain stem.[336] Due to its location and neural connections, the thalamus works for the brain stem in a way, but it is not limited to getting information from it. Instead, the thalamus represents the last bottleneck of information flow and provides a substrate to influence that flow.[337]

During the period of destruction, the inhabiter's enabled imprint strings will continue to create electromagnetic waves via two rotations of the four-step sequence, and an inhabiter will most definitely benefit by choosing the psyche altruistic. The inhabiter's beam of primordial consciousness widens the farther it travels from the focal point of dark energy, near the surface of the focal point it is narrowed. Thus, the extremely subtle conscious level of the layered electromagnetic radiation is illuminated by a narrowed beam of primordial consciousness. When the inhabiter's central region of its beam with coherent primordial intelligence is illuminating the space at the extremely subtle conscious level, the inhabiter finds peace while the computer brain's circuitry is malfunctioning and its programs are crashing. There are no electromagnetic waves with incorporated transformed neurotransmitter data present at the extremely subtle conscious level. For an inhabiter that has trained itself, this is a space where the inhabiter has the ability to break free from the constraints and traps of fear during this event.[338]

Because it will be the computer brain's large repertoire of neurotransmitter programs, especially those created by the brain stem that will set the associative rules followed by the meta-algorithm when processing furiously, there are certain neurotransmitters that might hinder an inhabiter to make beneficial choices during the period of destruction.

Glutamate: A neurotransmitter that when released enhances the synchronicity of the pulse propagating between two neurons, aiding in signal transmission and cross-talking between neurons in different circuits while the computer brain is computing for impermanence.[339] When ego-clinging imprint strings transform glutamate and the inhabiter experiences it as a feeling that precedes a sense impression or mental event of the psyche self-centered, the inhabiter will be confused and unable to keep track of its thoughts.

Norepinephrine: A neurotransmitter that will be released because the computer brain is mobilizing its circuitry to compute rapidly, and will be released at high levels during the computation for impermanence.

Enabled ego-clinging imprint strings will transform glutamate and norepinephrine concurrently, which may be experienced by the inhabiter as unpleasant feelings that precede physical sensations and mental events of the psyche self-centered, which include a rapid flow of thoughts and emotions that are anxiety provoking. Thus the inhabiter will experience stress through the pawn's body.

Each round of the game of life during the period of destruction is marked by a final conflagration uniquely experienced by each inhabiter.[340] What is important to consider is that the psyches contain electromagnetic waves of both sense impression and mental events. This means that the inhabiter will experience mental events as three-dimensional moving holograms but will also feel physical sensations. Which is to say that the period of destruction is an event that includes physical sensation that will be experienced differently by each inhabiter, dependent on its population of enabled imprint strings and choices. This makes this period a critical event in the game of life. If the inhabiter has an abundance of enabled ego-clinging imprint strings, the odds are stacked against the inhabiter that it will be able to avoid its habitual choice of the psyche self-centered. Instead, it is very likely that the inhabiter in this situation will experience the period of destruction as a nonlucid nightmare. This nightmare will consist of physical sensations with the intensity of the sense impressions corresponding to the number of ego-clinging imprint strings that are enabled, and the moving holograms perceived by the inhabiter may include seven increasingly consuming fires that will burn up the illusory visible world. [341]

When the normal matter computer brain computes its last computation and the inhabiter selects one of the two psyches to experience this last moment, the electromagnetic radiation of both psyches then dissolves into its focal point of dark energy. This causes a reverberation with a force that correlates directly with the inhabiter's last motivation. Notwithstanding, this reverberation enables and disables the inhabiter's imprint strings as it moves across the surface of the inhabiter's focal point of dark energy. When the reverberation reaches the inhabiter's imprint strings, it will determine the type of imprint strings and how many of them will be enabled and disabled. As the imprint strings are arranged on the inhabiter's focal point of dark energy as ego-clinging and non-ego-clinging pairs, a change in a magnetic charge through one type of enabled imprint string causes a magnetomotive force across the other that disables or enables the other type of imprint string. Specifically, an ego-clinging imprint string disables a non-ego-clinging imprint string, and the opposite mechanism is at work when non-ego-clinging imprint strings are enabled. The enabled imprint strings that are left, however, do not transmit alternating signals of magnetic charges that will be picked up by the normal matter computer brain, because upon performing its last computation, the computer brain has already started to degenerate. Instead, the universe's dark energy picks up the transmission of alternating signals of magnetic charges produced by enabled imprint strings because these imprint strings are located on the inhabiter's focal point of dark energy that is attached to the universe's substrate layer. When normal matter is no longer configured as a computer brain, it is freed from the interaction with the dark matter inhabiter and reverts to normal energy that is temporarily dispersed within the vast universe.

Inhabiters embark on their own odysseys of discovery and have a unique timeline when their interaction with normal matter will end. In order to liberate itself and normal matter from their interaction, the dark matter inhabiter must balance the equation of interaction by the disabling of its ego-clinging imprint strings. The inhabiter that has not managed to disable its ego-clinging imprint strings will participate in another round of the game of life, yet, in order to do so, the inhabiter needs normal energy temporality freed so as to form a physical framework for the interaction as a normal matter computer brain. This means that at any given moment in the universe, normal energy is in high demand. When there is not normal energy that is free and available, then the inhabiter's enabled imprint strings temporarily cease to oscillate, and the inhabiter is not embodied within a normal matter computer brain. Dark energy as the universe's substrate layer with omnipotence is aware of every event taking place simultaneously in the nonseparable universe, in part because the inhabiter's focal point of dark energy is attached to the universe's two-dimensional grid, which is the dark energy substrate layer. If there is not normal energy available, then the universe's dark energy substrate layer sends across the universe quantum signals to focal points of dark energy within inhabiters configurations, causing their enabled imprint strings to temporarily cease to oscillate.

- Nonmanifest state: The period that contains an intermediate void of created cognitive states. The nonmanifest state can be thought of as the period where the inhabiter exists between two universes. There is the illusory universe within the psyches perceived by the dark matter inhabiter while embodied within a normal matter computer brain and participating in a round of the game of life. The other universe is the real universe consisting of dark energy and the ambient normal energy that glides across the universe's substrate surface. The duration of the nonmanifest state is variable and depends on the availability of normal energy, as there is only 5 percent of the total energy composition that is normal energy in the universe. Normal energy in high demand most commonly occupies the universe in its visible form, normal matter. Until every inhabiter has completed its journey, normal energy will be drawn by the gravitational force of enabled imprint strings to embody inhabiters as normal matter computer brains.

When normal energy is unavailable, the inhabiter's imprint strings, though not oscillating, are attached to the inhabiter's perfectly black body, which is its focal point of dark energy. Enabled imprint strings demarcate the peripheral dimensions of focal points of dark energy and distinguish these focal points from the universe's dark energy substrate layer. Thus, inhabiters may exist attached to the universe's dark energy substrate layer with a surface of enabled imprint strings that are not oscillating and without a physical framework of normal matter. A focal point of dark energy with enabled imprint strings that are not oscillating is a special type of dark matter that does not produce a gravitational force, and the inhabiter with this configuration is known to have a *mental body*. An inhabiter configured with a mental body is in a bit of a quandary because in this situation the inhabiter must wait for a normal matter physical framework. However, during the period of nonmanifest state, the inhabiter will perceive a different sort of psyches, because electromagnetic radiation during the period of destruction dissolved into its focal point of dark energy that possesses primordial consciousness. The inhabiter's focal point of dark energy, having acquired the layered electromagnetic radiation, incurred the electromagnetic waves with forces created by the inhabiter's both types of imprint strings. Thus the positive forces created by the magnetic monopoles, enabled non-ego-clinging imprint strings were received as well as the negative forces created by the magnetic monopoles, enabled ego-clinging imprint strings. If a positive force creates the inhabiter's primary focus during the nonmanifest state, the inhabiter will experience this unique version of the psyche altruistic as a lucid dream with scenes of color and the perception that this is a positive sign. Thus the inhabiter will not react to this appearance; instead, hatred, anger, and passion dissolve.[342] With the law of cause and effect in force, the opposite scenario also takes place, and if a negative force creates the inhabiter's primary focus, the inhabiter will feel very alone and frightened while perceiving the psyche self-centered as a terrifying, unique nonlucid nightmare.

Within the inhabiter's focal point of dark energy, primordial consciousness is illuminated in all directions, but without normal matter surrounding the focal point of dark energy, a consistent, uniform temperature is not maintained. Thereby, the inhabiter's black body temperature actually increases, and primordial consciousness illuminating in all directions will appear to the inhabiter as if there is an emission of a significant amount of visible light.[343] From the inhabiter's perspective, it will appear as if the whole world is made of light[344] that will be accompanied by a corresponding initial sound of a "trrrrrr."[345] As the inhabiter's focal point of dark energy gets hotter, the inhabiter will perceive the color of red[346] right before intense darkness[347] and the inhabiter's experiencing of the void of pure consciousness.[348] At this time, the inhabiter will be free of the thoughts of created cognition, and if the forces absorbed by the inhabiter's focal point were predominantly positive, then the inhabiter will experience the void of its primordial consciousness as bliss and nonconceptuality. However, if the forces were predominately negative, then the inhabiter will experience the void of its primordial consciousness feeling frightened and completely alone.[349]

The nonmanifest state will be experienced by the inhabiter in this way until normal energy is available. At that time, the inhabiter's focal point of dark energy receives a signal from the universe's dark energy substrate layer that causes its enabled imprint strings to begin to oscillate and communicate a gravitation force to the normal energy close by. How normal matter folds around the inhabiter creates the shape and structure of the normal matter computer brain, and the inhabiter participates in another round of the game of life. This leads from the former normal matter computer brain's collapse to reexpansion.[350] As a singularity, the inhabiter with its enabled imprint strings reconfigures normal matter into a computer brain; this marks another interaction that will continue between these two energies, albeit with a new pawn and perhaps not the avatar of interaction that the inhabiter hoped for. There will be myriad unfamiliar neural connections that will replace those of previous circuitry, and each subsequent normal matter computer brain will have a finite lifetime and will collapse, only to reform.[351] Hence, the quantity of inhabiters in a given universe increases or decreases according to the availability of normal energy; the universe is cyclical though neither circular nor repetitive.[352] The fact that enabled imprint strings are not destroyed, only disabled and enabled, ensures that there is continuity between one illusory universe within the psyches and the next.

- Formation (another round in the game of life): The period where a dark matter inhabiter continues its interaction with normal matter and begins another round in the game of life, which begins with the sound of a thousand thunderclaps resounding simultaneously.[353] The inhabiter will perceive lights as enveloping it, and light rays that appear as sharp as a weapon.[354] These light rays are due to the gravitational forces of its enabled imprint strings interacting and guiding normal matter to configure another computer brain.[355] From the inhabiter's perspective, it will appear as if normal matter is sliding along straws of light back and forth, producing a halo effect of "soft hair"[356] made of the straws of light, until a new computer brain is configured. As the dark matter inhabiter again is embodied within a normal matter computer brain, the temperature of its focal point of dark energy begins to decrease. The normal matter computer brain begins to produce data, and the inhabiter's enabled imprint strings receive signals and transform them. If the inhabiter is embodied within a computer brain configured to produce data for a human psyche, the inhabiter first perception of the subtle conscious level of the layered electromagnetic radiation will be the full spectrum of the electromagnetic field as white, red, blue, green, and yellow.[357] Thus begins another round of the game of life with psyches that contain sense impressions and mental events as moving three-dimensional holograms of forms, textures, smells, tastes, and sounds.[358]

However, there is a paradoxical scenario in place in the universe that is reserved for an inhabiter nearing the completion of its journey, and this inhabiter is called an *enlightened inhabiter.* An enlightened inhabiter has disabled all of its ego-clinging imprint strings, except it chooses to keep one of these strings enabled. Accordingly, this inhabiter is one that could choose to end its odyssey of discovery and cease to live a life of conditioned existence to become a part of the universe's dark energy substrate layer, but it chooses not to. As follows, there are inhabiters on their way to becoming an enlightened inhabiter, while others are on their way to disabling most of their non-ego-clinging imprint strings. The enlightened inhabiter is no different from other inhabiters in this way: the enlightened inhabiter began its journey interacting with normal matter with an equal number of enabled ego-clinging and non-ego-clinging imprint strings. Each inhabiter is dark energy that chose to transform itself to a focal point of dark matter. Albeit, dark energy understood the ramifications before beginning its interaction with normal energy and did not gamble or play dice when designing the variables to the equation of its interaction.[359] Instead, dark energy threw itself wholeheartedly into its interaction as a dark matter inhabiter and designed the equation in a way that allowed the normal matter computer brain to have the opportunity to react according to its nature. This means the first choice each dark matter inhabiter makes when beginning its first round in the game of life while embodied within a normal matter computer brain will be to choose the psyche self-centered. This also means that there is no easy exit from the game of life, as this initial choice causes each inhabiter to become immersed in the cognitive states of the psyche self-centered, though the duration of the immersive experience will vary between inhabiters. What differs between an enlightened inhabiter and other inhabiters is that after inhabiting many normal matter computer brains while participating in untold rounds of the game of life and perceiving pawns of many different shapes, colors, size, and species, the enlightened inhabiter trained itself to be a skilled participant. Thereby, the enlightened inhabiter transformed itself into a mental detective with the means by which to discern evidence in three-dimensional holograms and consistently choose the psyche altruistic that contained cognitive states for alternative motivations and beliefs. Correspondingly, with each round of the game of life, this inhabiter was embodied by normal matter computer brains with increased capacities, and the enlightened inhabiter reprogrammed circuitry and brought its enabled imprint strings and focal point of dark energy with it to each new game.

Thus the enlightened inhabiter was able to disable all of its ego-clinging imprint strings but chose to keep one enabled because it was no longer lost in the nonlucid dream or nonlucid nightmare of the psyche self-centered. Instead, this inhabiter obtained wisdom by choosing the psyche altruistic that radically changed its perspective and the way it participated in the game of life. Thereby, the enlightened inhabiter grasped the deeper meaning to its existence and the complexity of the nonseparable universe in a way that motivated it with the sincere desire to benefit all inhabiters and normal matter configured as computer brains. The universe's dark energy substrate layer facilitates entanglement but does not engage in interactions directly with normal matter. Dark energy as the universe's substrate layer does not punish any inhabiter; rather, it facilitates the dispersal of quantum information in a way that coincides with each inhabiter's motivations, allowing them to self-actualize their choices. Ergo, if an inhabiter does not have at least a single enabled ego-clinging imprint string, this would mean that its odyssey of discovery has ended. In this scenario, the inhabiter's population of enabled non-ego-clinging imprint strings are thereby absorbed by its focal point of dark energy. However, only an inhabiter embodied within a normal matter computer brain can alter circuitry and communicate with other inhabiters via entanglement. The enlightened inhabiter remains embodied within a normal matter computer brain to self-actualize its choice to be of benefit to all the energies present in the nonseparable universe but in a way that differs from the universe's dark energy substrate layer. While the enlightened inhabiter is embodied, the universe's dark energy substrate layer passes signals from this inhabiter to other inhabiters spread throughout the nonseparable universe, no matter how far apart they are. Nonetheless, the signals sent by an enlightened inhabiter differ from those of other inhabiters because the enlightened inhabiter and the universe's dark energy substrate layer work together and with the same purpose, which is to spread love and compassion throughout the nonseparable universe in two profoundly different ways. The universe's dark energy substrate layer nonconceptually facilitates entanglement without borders or controls. The enlightened inhabiter's approach, however, is to encourage inhabiters to choose the altruistic cognitive states of the psyche altruistic and thereby provide themselves with the means by which to disable their ego-clinging imprint strings. Having transformed itself to an enlightened inhabiter, life of conditioned existence is not experienced in the same way as it is for other inhabiters. This inhabiter is lucidly aware, and there is no ambiguity in its perception; nor is an enlightened inhabiter fooled by illusory appearances.

However, there are limits as to how much an enlightened inhabiter can help another inhabiter; it depends on the receiving inhabiter's choices and what species of psyche is being perceived. Furthermore, when an inhabiter is perceiving a human psyche and the central region of its beam of primordial consciousness is illuminating the subtle conscious level of the layered electromagnetic radiation, then the inhabiter will formulate its perceptions with holograms of cultural influences that will help shape its beliefs. Compounding the complexity is that the enlightened inhabiter's signals will be incorporated contextually into the psyche self-centered and the psyche altruistic in a way that produces contrasting viewpoints. Albeit, the enlightened inhabiter has unlimited creative power. When its signals get incorporated into the electromagnetic waves created by enabled non-ego-clinging imprint strings that contain transformed neurotransmitter data and are illuminated by the central region of the inhabiter's beam of primordial consciousness, three categories of holograms emerge at the subtle conscious level: a metaphorical role model, companions, and media of the scientific and (or) spiritual. Through these categories, the inhabiter will be guided in cognitive experiences that include discipline, concentration, and knowledge,[360] and these will be presented within scenes of the psyche altruistic in ways that will be easy for the inhabiter to assimilate[361] if studied and reflected upon. Because it is the inhabiter that must transform itself from one that builds its beliefs based on appearances to another that formulates its beliefs through understanding, the only constant in the universe, which is causality.[362] When the inhabiter questions why things appear as they do and diligently searches for answers within these categories, a thought emerges that there is something wrong with its perception of the world. It is this thought that brings the inhabiter to seek and follow guidance. If the inhabiter remains open to what it learns, then it begins to grasp the relative and infinite dimensions of its reality, and with a widened perspective, the inhabiter has the power to change its inaccurate beliefs and redefine what is real.[363]

The metaphorical role model produced by quantum signals sent by the enlightened inhabiter creates a being either living, or the inhabiter when thinking with a human psyche might perceive this character as having lived in the past. Whichever the case might be, the metaphorical role model is meant to guide the inhabiter into the metaphysical realm of realty via spiritual and authentic contemplative/meditative training. The inhabiter will be drawn to the metaphorical role model because of the connection it feels to the complex set of attitudes and self-regulatory behaviors that the inhabiter can identify with and use as behavioral guides. Yet, without knowing it, the inhabiter is experiencing the identity principle, because when it follows the metaphorical role model, it is actually connecting with its primordial consciousness in a very meaningful way. The psyche altruistic, therefore, includes ordinary people, but the character that was created with an enlightened inhabiter's signals is in no way ordinary.[364] This character represents pure consciousness of an enlightened inhabiter that made a decision to remain embodied within a normal matter computer brain to be of benefit. The inhabiter may read of the metaphorical role model levitating, walking on water, flying in the air, passing through rock, and performing numerous other feats.[365] Be that as it may, the greatest powers this character possesses are its compassion, empathy, enlightened intelligence, and creativity. Thus it is power with grace that extends beyond[431] conditioned existence and reaches to the nonseparable universe.[366]

The means by which an inhabiter is able to distinguish the metaphorical role model as being in the psyche altruistic is that the character's activities are for the sake of other living beings.[367] Nothing this character does is to become famous, and its accomplishments are purely motivated by the desire to benefit others.[368] Accordingly, the metaphorical role model assumes a humble posture, even though this character is endowed with tremendous qualities that lead to altruistic activities.[369] Further clues are the metaphorical role model's qualifications as being an expert in spiritual texts, and the character has direct experience of what it is teaching. Since the very purpose of the metaphorical role model is to teach the inhabiter the means by which to attain mental stability, a teacher of such knowledge could not help the inhabiter tame its mind without having first having tamed its own. The metaphorical role model is the way in which an enlightened inhabiter can share with the inhabiter its past experiences participating in rounds of the game of life, and how this inhabiter transformed itself through morality, meditative training, and wisdom, which is what the inhabiter must do.

- Morality: Behavior clue, displayed in the words and actions of the pawn. When the inhabiter detects this behavior and selects the psyche altruistic, unpleasant and pleasant feelings that precede physical sensations and mental events are controlled by disinterest. Thus these feelings lose their power to pull the inhabiter into unfit actions and the selection of the psyche self-centered.[370]
- Study/reflection/meditative concentration: Behavior clues, where study and reflection will cut through the inhabiter's more gross misconceptions, but the subtler ones can only be dispelled by

meditation and by integration of the state of pure awareness that arises from sustained calm and profound insight.[371] One-pointed meditative concentration allows the inhabiter to overcome its excitement while perceiving moving holograms that serve as a distraction, when the central region of the inhabiter's beam of primordial consciousness is illuminating the electromagnetic waves at the subtle conscious level.[372]

- Wisdom: A quality that the inhabiter acquires while thinking with the cognitive states of the psyche altruistic, and it produces the inhabiter's disbelief that the pawn is "self." With wisdom, the inhabiter possesses a relative but accurate viewpoint that allows the inhabiter to comprehend that what is perceived at the subtle conscious level of the layered electromagnetic radiation has no inherent existence. This will serve to pacify the inhabiter from responding to cognitive states while perceiving the psyche self-centered that has unpleasant feelings that precede counterproductive emotions. Therefore, the inhabiter will be less apt to make a hasty response; instead, the inhabiter will learn to patiently wait until it perceives clues that indicate the psyche altruistic, which might include sacred scriptures. Thus, with wisdom, the inhabiter makes its selection and chooses the psyche altruistic, and the inhabiter thinks with cognitive states that produce thoughtful reasoning.[373]

The metaphorical role model will guide the inhabiter to an understanding of the deeper dimensions of its reality through altruistic themes, as a foundation for understanding and skill development. This allows the inhabiter to explore the varied ramifications of its decisions and to experience empathy for other characters.[374] As the inhabiter makes successful attempts and discerns its motivations while picking up behavior clues in the pawn's words and actions, then selects the psyche altruistic, the inhabiter is rewarded through accurate perception and positive emotions. Unsuccessful attempts and the choice of the psyche self-centered are followed up with more clues and another opportunity for the inhabiter to select the alternative perspective. In this way, each scene of the psyche altruistic at the subtle conscious level is a training session that acts as a measure of the inhabiter's understanding and may include themes of anguish surrounding life and death[375] or thrilling temptations of reward if feelings rather than behavior clues are followed. Whatever they may be, the inhabiter is guided by pure consciousness that teaches the inhabiter an essential principal:[376] without picking or choosing, every animate or inanimate object perceived serves as the path for the inhabiter to obtain wisdom and disable its ego-clinging imprint strings.

Rather than trying to evaluate every aspect of a potential action or decision and becoming overwhelmed, the inhabiter can simply ask itself, "What would my metaphorical role model do in this situation?" By asking this simple question, the inhabiter chooses behaviors that will, in fact, allow the inhabiter to achieve a goal of which it cannot yet remember, during the period when the inhabiter was immersed in the cognitive states of the psyche self-centered. The inhabiter's goal was self-actualized by a choice it made an unknown time ago when part of the omnipotent dark energy that transformed itself from the universe's substrate layer to a vulnerable focal point of dark matter so that it might learn the most beneficial way to coexist with normal energy. Now the inhabiter has to understand why it made the choice but must do so while thinking with the cognitive states of the psyche altruistic and disable its ego-clinging imprint strings before the normal matter computer brain computes for impermanence.[377] Having to choose while thinking with cognitive states is a rule to the game of life that leaves the inhabiter vulnerable. Because, when the inhabiter chooses the psyche self-centered, life in the universe is portrayed inaccurately, and this causes the inhabiter's habitual selection that results in ignorance being the inhabiter's modus operandi while immersed. Wisdom, however, sets the inhabiter back on its path when transformed as a foothold for action and the choice of the psyche altruistic. Thereby the inhabiter has the skill to detect clues that reveal when it has stumbled with rage, hate, jealousy, and pride into the darkened dimensions of the psyche self-centered. Thus the inhabiter possesses the mean by which to transform itself, and the fruition of the inhabiter's efforts are actualized in a way that normal matter benefits by the balancing of the equation of interaction and the disabling of ego-clinging imprint strings.[378]

When an enlightened inhabiter's quantum signals are dispersed throughout the nonseparable universe and received by inhabiters' focal point of dark energy, what might emerge as three-dimensional holograms within the psyche altruistic are endless, thus fulfilling the endless possible needs of inhabiters lost in a nonlucid dream or nonlucid nightmare of the psyche self-centered. Accordingly, an enlightened inhabiter having experienced conditioned existence many times during copious rounds of the game of life is aware that if an inhabiter experiences sickness, painful physical sensations such as extreme temperatures and hunger, these cause potent distractions and will hinder an inhabiter's ability to disable its ego-clinging imprint strings through beneficial choices. For this reason, the enlightened inhabiter might send to inhabiters within disparately configured normal matter computer brains signals that create within their human and nonhuman psyches clothes, medicines, shelter, and food. Perhaps the quantum signals might create what inhabiters, while thinking with diverse human cognitive states, perceive as temples, mosques, churches, synagogues, sacred places in nature, symbolic spiritual forms, songs, and prayers.

What is not revealed to inhabiters in three-dimensional holograms is the duration in which an enlightened inhabiter remains embodied within a normal matter computer brain, as this information is only known by the universe's dark energy substrate layer and an enlightened inhabiter. What is not a mystery is the enlightened inhabiter's motivations that are deeply rooted in its unfaltering commitment to remaining in a conditioned state of existence for as long as it takes, until each inhabiter has the opportunity to complete its odyssey of discovery, as this would benefit the nonseparable universe.

Key Point

- It is enabled imprint strings that ultimately determine what happens to an inhabiter and normal matter at brain death.

The nonseparable universe is where inhabiters experience the periods of formation, continuity, and destruction and where there are enlightened inhabiters that are spread throughout the universe for benefit.

Formation: Marks the beginning of an initial interaction between a dark matter inhabiter and a normal matter computer brain, or represents an inhabiter's continued interaction with normal matter and the start of another round in the game of life.

Continuity: The period where the dark matter inhabiter interacts with normal matter while embodied by a computer brain.

Destruction: The period that starts when the normal matter begins its computation for impermanence and ends when the inhabiter is no longer embodied by a computer brain.

Enlightened inhabiter: An inhabiter that has the ability to disable all of its ego-clinging imprint strings but chooses to keep one enabled to be of benefit by sending quantum signals to inhabiters spread throughout the nonseparable universe.

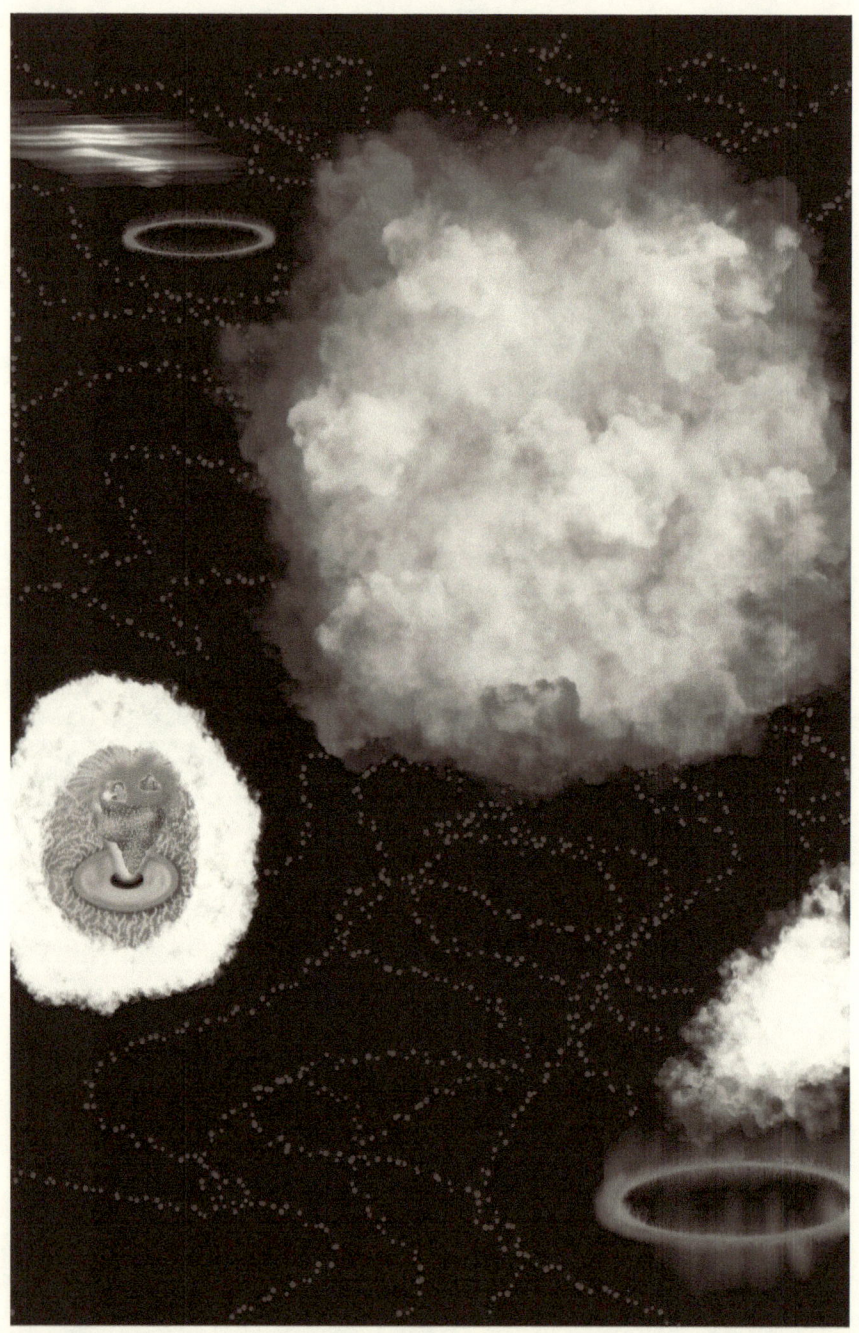

Formation (bottom right), continuity (midleft), destruction (upper left), enlightened inhabiter (upper right) sending quantum signals.

Nonmanifest state: The period that contains an intermediate void of created cognitive states and that is variable in length, dependent on the availability of normal energy. When normal energy is unavailable, the inhabiter's imprint strings cease to oscillate but are attached to the inhabiter's focal point of dark energy. Enabled imprint strings demarcate the peripheral dimensions of focal points of dark energy and distinguish these focal points from the universe's dark energy substrate layer. Thus, inhabiters may exist attached to the universe's dark energy substrate layer with a surface of enabled imprint strings that are not oscillating and without a physical framework of normal matter. A focal point of dark energy with enabled imprint strings that are not oscillating is a special type of dark matter that does not produce a gravitational force, and the inhabiter with this configuration is known to have a mental body. The layered electromagnetic radiation and electromagnetic waves with forces created by the inhabiter's both types of imprint strings dissolved into the inhabiter's focal point of dark energy.

If a positive force creates the inhabiter's primary focus during the nonmanifest state, the inhabiter will experience this unique version of the psyche altruistic as a lucid dream with scenes of color and the perception that this is a positive sign. Thus the inhabiter will not react to this appearance, but instead, hatred, anger, and passion dissolve.

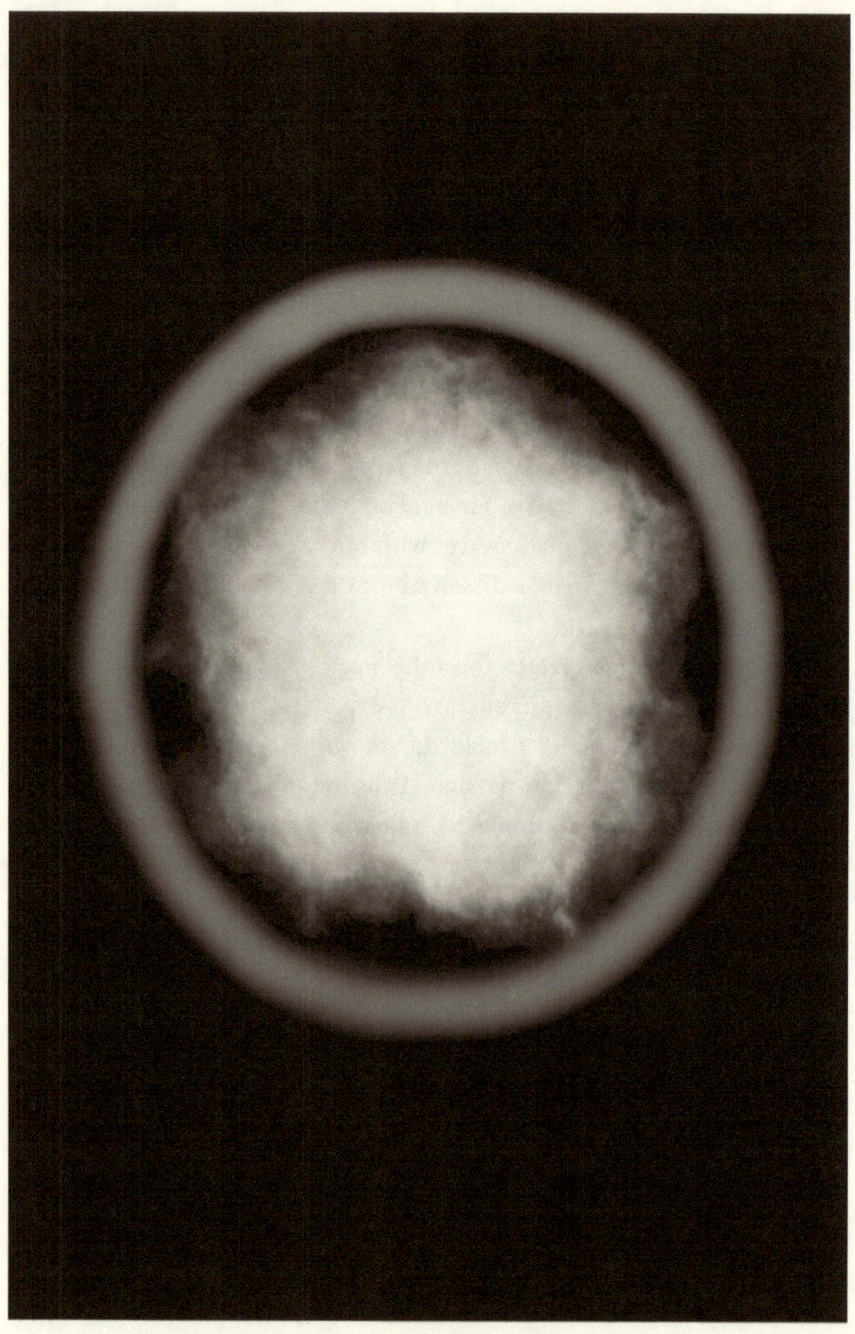

Inhabiter with enabled imprint strings that are not oscillating
experiencing a unique lucid dream of the psyche altruistic.

With the law of cause and effect in force, the opposite scenario also takes place, and if a negative force creates the inhabiter's primary focus, the inhabiter will feel very alone and frightened while perceiving the psyche self-centered as a terrifying, unique nonlucid nightmare.

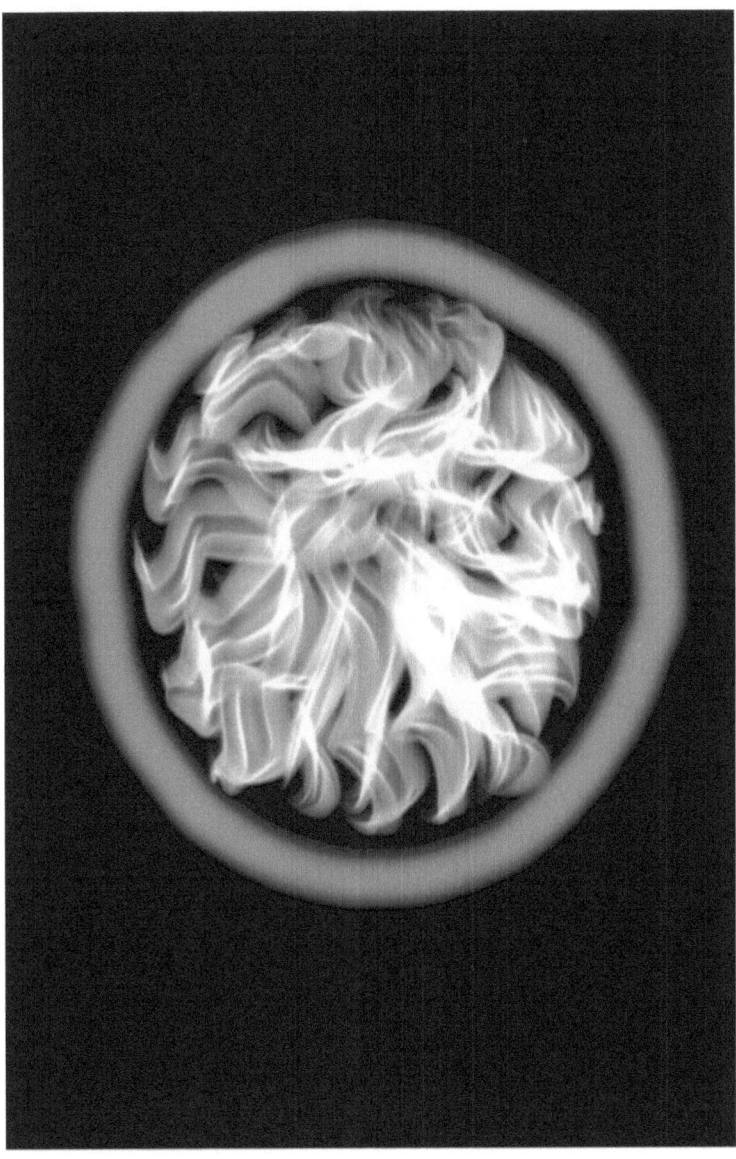

Inhabiter with enabled imprint strings that are not oscillating experiencing a unique nonlucid nightmare of the psyche self-centered.

Reminders before Reading Conclusion

Variables Built into the Equation of Interaction

- The inhabiter's imprint string population can increase in number as needed to fulfill the demand, and its focal point of dark energy will simply morph and make more.
- To prevent a dark matter inhabiter from becoming completely immersed with normal matter without the means by which to be enduringly freed from its interaction, a single non-ego-clinging imprint string was earmarked as a safety measure. Thus the inhabiter's focal point of dark energy prevents this one imprint string from being disabled.

Four Periods to the Game of Life

- formation: The period marked by the beginning of an interaction between a dark matter inhabiter and a normal matter computer brain.
- continuity: The period in which the dark matter inhabiter interacts with normal matter while embodied within a normal matter computer brain. During this period, the inhabiter balances or unbalances the equation of interaction via its choices.
- destruction: The period that begins when normal matter starts to compute for impermanence and ends when the inhabiter is no longer embodied by a computer brain.
- nonmanifest state: The period that contains an intermediate void of created cognitive states and that is variable in length, dependent on the availability of normal energy.

Miscellaneous Topics

- mental body: A phenomenon that takes place when the inhabiter is not embodied within the physical framework of a normal matter computer brain and its enabled imprint strings are not oscillating.
- enlightened inhabiter: An inhabiter that has the ability to disable all of its ego-clinging imprint strings but chooses to keep one enabled to be of benefit to the nonseparable universe.
- metaphorical role model: A living being that emerges at the subtle conscious level of the layered electromagnetic radiation when an enlightened inhabiter's signals are incorporated into the electromagnetic waves created by enabled non-ego-clinging imprint strings.

CONCLUSION

In absolute truth, there are not living beings that are born, and there are no living beings that die except within the psyche, as the total sum of the universe's mass and energy does not change.[379] Therefore, nothing can start to exist or cease to exist, and where there is no violation of the principle of the conservation of mass-energy, there are only transformations.[380] A phenomenon's apparent properties derive from the complete set of phenomena, primordial consciousness included.[381] Nothing would emerge within the psyches and be discernible to inhabiters be it not for the fact that they had within their configuration a focal point of dark energy with primordial consciousness. Though phenomena appear as three-dimensional holograms within the psyches, what illuminates electromagnetic waves with incorporated neurotransmitter data is primordial consciousness that is empty of created cognitive states.[382] The fact that primordial consciousness is empty of created cognitive states but detects transformed neurotransmitter data and reacts by illuminating electromagnetic waves is not just the nature of phenomena; primordial emptiness is also the potential that allows the propagation of an infinite variety of phenomena to appear.[383] Primordial consciousness fashions reality, and reality fashions consciousness.[384] An illustrative example would be the concept of time that exists only within the psyches but is the reference by which the inhabiter contextualizes its mental experience. Each instant is a perpetual end and beginning because of the basic impermanence of the phenomena produced by the laws of cause and effect. Notwithstanding due to the subtle impermanence of phenomena, no measurement can be truly instantaneous, considering a measurement occurs in time and cannot be absolutely accurate, for conditions are constantly changing.[385] In terms of absolute truth, all past, present, and future events are identical in that they have no intrinsic existence, yet they emerge as three-dimensional holograms at the subtle conscious level of the layered electromagnetic radiation and appear within one of the psyches. Thus, events have no real end or beginning, and if nothing is really produced, there is no need to look for an end.[386] Reality is unbelievably complex and underlies illusions such as that of a living being existing because of a single male and female entity. Instead, a novelty such as the birth of a living being that appears only within the psyches comes about from collaboration rather than pure chance or explained by a limited number of causes.[387] Since the beginning of time, inhabiters have been communicating with one another via the phenomena of entanglement during many rounds while participating in the game of life. This means that there exists an undetermined number of variables that have a cause and effect relationship, within an infinite number of equations of interactions taking place within the nonseparable universe.[388]

Dark matter inhabiters and normal matter computer brains are intimately connected, and the universe's dark energy designed the variables to the equation of interaction this way. A dark matter inhabiter interacts with normal matter but within the physical framework of a computer brain that acts as a firewall, thereby containing the experience.[389] The universe's dark energy transformed itself into a dark matter inhabiter while fully aware that its interaction with normal matter is unstable and the equation of interaction can become critically unbalanced. When interacting with each other, inhabiters and normal matter computer brains are unpredictable, and the many versions of the psyches perceived by inhabiters are impermanent. The chaotic life of conditioned existence is contained, however, so that it does not affect the transcendent, real universe.

Though appearances with the psyche self-centered and the psyche altruistic are many, they are deceptive, because in reality, there were always only two energies interacting. While interacting, neither the dark matter inhabiter nor the normal matter computer brain is free. When the equation of interaction becomes unbalanced by inhabiters choosing the psyche self-centered, life, love, and compassion have restrictions. Only the inhabiter—not the computer brain—will perceive characters as having form, and it is the inhabiter that will assign meaning to the characters based on their relationship to the pawn. Through these characters, the immersed inhabiter will identify its place in the world of normal matter and will experience emotions and learn about impermanence as it sees characters born and die. This means all characters, including enemies and nonhuman life forms, are important, because without them, the inhabiter could not progress on its path to reemergence. While there are many characters perceived by each inhabiter at the subtle conscious level of the layered electromagnetic radiation, there is only one pawn that is the avatar that allows the inhabiter to directly experience its interaction with the normal matter computer brain. If the inhabiter does not understand its role in the creation of its cognition, it will be an unskilled participant in the game of life while experiencing conditioned existence. Both the psyche self-centered and the psyche altruistic will display the pawn that has family and friends, and the pawn may be perceived by the inhabiter as conforming to societal constraints by working, having a Social Security number, and paying taxes.[390] Albeit, the inhabiter perceiving itself as the pawn has a profound attachment to only those to whom it feels connected, yet the word itself is nothing but the inhabiter's interpretation of neural synchrony generated by a normal matter computer brain that does not feel the same connection that the word implies.[391] What happens to the dark matter inhabiter that immerses itself in the cognitive states of the psyche self-centered and why it habitually chooses this psyche to hold on to synaptic connections[392] that create living beings, places, and things is ultimately harmful to itself and normal matter.

Although there are no actual countries or a planet where things are owned and where weapons of mass destruction exist in the nonseparable universe, there are inhabiters communicating via entanglement. For an undisclosed but variable amount of time, immersed inhabiters forget their purpose in entering the interaction with normal matter and lose their ability to perceive without duality. Instead, they redefine their life purpose, making the psyche self-centered their modus operandi, and to varying degrees, pawns exhibit predatory and narcissistic behavior. The inhabiter, misapprehending its conditioned existence, has not understood the very real threat that exists because the universe is nonseparable and has a dark energy substrate layer that facilitates entanglement. However, it is not the nonseparable universe or the communication between inhabiters that causes the threat; rather, it is the abundance of inhabiters that experience

191

conditioned existence lost in nonlucid dreams and nightmares of the psyche self-centered. Thereby, dark matter inhabiters off-balance the equation while interacting with normal matter computer brains; normal matter has no free choice made with cognitive states. Choice is a luxury afforded to the inhabiter, and only when there are sufficient numbers of its enabled non-ego-clinging imprint strings does the inhabiter have the power to control its perception. The degree to which its non-ego-clinging imprint strings are enabled corresponds to the inhabiter's ability to freely choose its actions without being influenced by the computer brain directing itself. Free, however, does not mean unconstrained, since the inhabiter's actions are free but tallied[393] by its enabled imprint strings. Inhabiters that are not beneficially communicating among themselves and due to entanglement might culminate to produce a group nightmare of the psyche self-centered of epic proportions.

The dark matter inhabiter and the normal matter computer brain are together, but the inhabiter is like a guest,[394] and the vessel of the computer brain like a hotel[395] in which that guest will only be making a short stay. Whereas the inhabiter can use mathematics to understand the fabric of the computer brain's circuitry, the exact time when normal matter computes for impermanence is not revealed. If inhabiters witness the same nightmare of the psyche self-centered that consists of a world war and nuclear annihilation, this means that it is not quantum signals of love and compassion spread throughout the nonseparable universe by the dark energy substrate layer. Instead, it is ignorance and suffering of incomprehensible dimensions that will be experienced by inhabiters while thinking with cognitive states. The universe's dark energy substrate layer is not facilitating entanglement in this way to delve out punishment; rather, this is the sacrifice and risk it took by transforming itself to dark matter inhabiters. Each inhabiter's configuration includes a focal point of dark energy imbued with primordial consciousness, which is pure awareness that does not think with cognitive states but is self-illuminating, like the flame of a lamp that illuminates other objects but does not need to illuminate itself, for light is its very nature.[396] While each inhabiter is on their unique journey, interacting with normal matter while embodied within computer brains, inhabiters will think with many diverse cognitive states that will arise and pass away. The primordial consciousness of dark energy is behind every experience in the inhabiter's life of conditioned existence, and this intelligence determines the way the inhabiter sees the illusory world while interacting with normal matter. Thus the inhabiter's primordial consciousness belongs to the realm of experience and cannot be accurately described with a third-person perspective that includes he, she, it, him, her, its, they, them, their, or theirs.[397] The inhabiter need not stand in incomprehension of the primordial consciousness within its configuration;[398] rather, understand that the present instant of its created conscious states has been triggered by former reactions of its primordial consciousness. With each of the inhabiter's thoughts made with the cognitive states of the psyche self-centered, the inhabiter reifies what it experiences and makes a false distinction between consciousness and the illusory world perceived.[399] Albeit, the inhabiter has within its configuration primordial consciousness that does not begin or end and is derived from its focal point of dark energy that is attached to the universe's dark energy substrate layer. This means the primordial nature of phenomena lies above and beyond notions of subject and object, of space and time,[400] and primordial consciousness cannot be isolated from the rest of reality; rather, it is intimately bound up with the whole of reality.[401] Dark matter inhabiters do not actually die; nor does normal matter. Rather, it is temporarily freed and residing in the universe as normal energy. However, contemplate the immediate demand for physical frameworks if so many inhabiters and normal matter computer brains abruptly end their interaction with a period of destruction of such monumental proportions of a group nightmare of world war and nuclear annihilation. There will be many inhabiters left for some time as a special type of dark matter and configured with a mental body during a prolonged period of the nonmanifest state. Only those inhabiters that have a single ego-clinging imprint string enabled during this destruction and could manage to disable it have a choice. They can disable the ego-clinging imprint string and become a part of the universe's dark energy substrate layer or retain it and become an enlightened inhabiter. Their decision will be based on what is most beneficial for the nonseparable universe. The rest of the inhabiters will be repaying their debt based on the number of their ego-clinging imprint strings enabled.

There are two great sources of fear[402] for the inhabiter: the moment the

193

dark matter inhabiter is "born" into conditioned existence with a surface of two types of fluctuating enabled imprint strings and the moment of brain death.[403] The inhabiter began its very first round of the game of life with a purposeful choice and selected the psyche self-centered, thereby exposing itself to becoming immersed in a nonlucid dream or nonlucid nightmare. During the period of continuity right up to the completion of destruction, the inhabiter made choices that balanced and unbalanced the equation of interaction. Thus, it is foolish for the inhabiter to say to itself that it is ready to leave its interaction with normal matter and that it has completed its odyssey of discovery without having done any work to disable its ego-clinging imprint strings. In other words, for as long as the inhabiter allowed itself to be immersed, the inhabiter has correspondingly enabled its ego-clinging imprint and thereby incurred a debt. Since enabled ego-clinging imprint strings do not simply disappear until the inhabiter manages to disable them, the inhabiter will repay the debt through the cognitive states of the psyche self-centered, experienced as a nonlucid dream or nonlucid nightmare. This is what causes the inhabiter to suffer.

Every impression the inhabiter had of an outer world and its living inhabitants had beginnings derived from impermanent programing.[404] That is how it works in the game of life.[405] If inhabiters cannot liberate themselves and normal matter from their interaction by balancing the equation, there are some bits of information that inhabiters keep in the form of their enabled imprint strings and the electromagnetic radiation that dissolved into their focal point of dark energy.[406] But the data for material possessions, loved ones, enemies, and the phenomenal world itself are lost during the period of destruction. Upon performing its last computation, the computer brain degenerates. The pawn, having been the avatar representing the inhabiter's direct interaction with normal matter, disappears with the computer brain that no longer is producing data and everything that surrounded the pawn as the phenomenal world vanishes.[407]

As follows, there are innumerable variables in an unknown number of equations of interactions within a nonseparable universe that consists of infinite relationships that condition each other mutually. The way an inhabiter perceives a particular living being comes about through many causes and conditions. The pawn and other living beings have no actual reality, as they exist only within the psyche self-centered and the psyche altruistic. Accordingly, a character's characteristics do not belong to them intrinsically; rather, the inhabiter's perceptions will be based on the cognitive states of the psyche chosen by the inhabiter. Living beings are created via quantum signals distributed by the universe's dark energy substrate layer, received by enabled imprint strings when they are diligently creating their mental events within their respective psyches. The information will be incorporated contextually by enabled imprint strings into both the psyche self-centered and the psyche altruistic in a way that complements mental events and make sense to the inhabiter. Then enabled imprint strings send signals to the computer brain, and in this way, the signals sent by the universe's dark energy substrate layer become a part of the normal matter computer brain's processing history. When the computer brain's meta-algorithm is processing for the core-I character or autobiographical-I character, there is no certainty how the living being will appear in either psyche.

When selecting the psyche self-centered, the inhabiter's perspective of reality is that appearances within the illusory world are separate from and external to itself as the pawn. Erroneously identifying itself as the pawn, what the inhabiter experiences becomes a very personal misconception of the meta-algorithm's methodology in using its I character as a central focus for executing programming. When the computer brain is directing itself, there is circuitry activated all over the place; correspondingly, the claustra are processing neural synchrony as electrical patterns and are highly reactive. What the inhabiter perceives in the psyche self-centered, experienced as a nonlucid dream or nonlucid nightmare, as obstacles or evil influences, such as ghosts, werewolves, aliens, or enemies, are not at all entities outside the subtle conscious level of the layered electromagnetic radiation. To the meta-algorithm, it is synchrony that fits, but to the inhabiter, it appears as if apparitions within its world are doing something they are not supposed to be doing.[408]

A normal matter computer brain does not think in extremes of bad or good; it detects only degrees of synchrony. No personal attacks directed at the inhabiter are sequenced for by normal matter, because the computer brain does not "see" anything and does not perceive the inhabiter. Only the inhabiter that has chosen the psyche self-centered sees characters with indifference or as loved ones, enemies, or predators. There are times when the inhabiter will struggle with the cognitive states of the psyche self-centered with hatred, pride, and jealousy.[409] These cognitive constructs exert their influence on the inhabiter's perception and poison it.[410] Awareness devoid of accurate information produces the inhabiter's impressions of characters and situations as being adverse, deplorable, and troublesome, and it responds as if it is being attacked. Each agent sequenced for by the meta-algorithm when transformed by enabled ego-clinging imprint strings creates a new effect[411] for the inhabiter at the subtle conscious level, depicting the pawn seeking food, fluids, and reproductive opportunities. It is the feelings of pleasant, unpleasant and neutral that motivate these inhabiters and with feeling the inhabiter inappropriately concludes that the pawn's needs are more important than those of other characters, hence cause and effect.[412] However, the meta-algorithm is not thinking this but is rather computing according to its normal matter nature. Normal matter does not "see" the agents the inhabiter perceives as being beautiful or ugly. The computer brain does not care if it is praised or insulted and has no thoughts of getting something in return for generosity, as these ideas are exclusive to an inhabiter having chosen the psyche self-centered.[413] Having no attachment whatsoever, the meta-algorithm follows rules that dictate how it will continuously sequence for agents. As it does this, enabled imprint strings transform this data that when illuminated by primordial consciousness generate the innumerable chains of the inhabiter's habitual reactions of attachment and aversion to things and the belief that the virtual world has a true and tangible existence.[414]

The root of the inhabiter's wanderings in the cognitive states of the psyche self-centered is the cycle of enabling rather than disabling its ego-clinging imprint strings. This traps the inhabiter in a relative perception of the world, preventing it from distinguishing between the way things appear and the way things are.[415] It is the inhabiter's relative perceptions that cause it to cling to people, places, and things—and attachment to these will be the inhabiter's downfall.[416] The inhabiter will covet forms and be driven by comforts, and as its ego-clinging imprint strings are enabled, it perpetuates the cycle of becoming like a moth, which, when attracted to light, gets burned by the flame.[417] There is an unequivocal relationship between the inhabiter's fervor in choosing the psyche self-centered and the anxiety the inhabiter experiences while discerning which impermanent scenes will bring conditions that create its happiness and avoiding those that bring its suffering, motivated by its hope and fear. The inhabiter believes itself to be the I character so blindly that instead of using the pawn's words and actions as behavioral clues and the means by which to avoid making a nonbeneficial choice, the inhabiter willingly enables its ego-clinging imprint strings. The inhabiter enables its ego-clinging imprint strings while witnessing the pawn sacrificing the lives of other living beings to satiate what the inhabiter thinks it needs. The mistaken belief that the pawn is "self" and that this illusion comes first is the ongoing theme experienced by the inhabiter while witnessing the pawn eating good food, wearing nice clothes,[418] in the nonlucid dream of the psyche self-centered. However much enjoyment the inhabiter may derive from perceiving pleasures does not negate that they are nothing more than three-dimensional holograms at the subtle conscious level of the layered electromagnetic radiation.[419] Yet a resounding and recurrent theme of "more" plays out within scenes of the psyche self-centered, and the inhabiter perceiving the pawn as itself is never satisfied. Instead, the inhabiter craves creature comforts, wealth, and status and desires to multiply itself. The most dangerous illusion for the inhabiter having chosen psyche self-centered is not that everything appears real; it is the inhabiter's mistaken assumption that it is in control of its life. By misidentifying itself as the pawn, the inhabiter gives up its personal control and the ability to steer the complex vessel of normal matter toward a beneficial interaction, as the inhabiter does not understand the rules to the game of life. The old adage of "what goes around comes around" will be true as the inhabiter will participate in many rounds of the game of life while perceiving different pawns and inhabiting varied normal matter computer brains, sometimes with very limited data generation and programming capacity.

Just as lighting a candle can reveal a hidden pathway, the inhabiter's belief that there actually exists enemies and insignificant living beings can be equally revelatory. This assumption is a clue that if the inhabiter picks up on it clearly reveals that its primordial beam with a central region of coherent intelligence is illuminating the electromagnetic waves of the psyche self-centered. If the inhabiter selects this psyche, then it has weakened its ability to participate in the game of life as a skillful participant but has incurred a debt and depleted its account at the bank of its focal point of dark energy by enabling its ego-clinging imprint strings. The psyche self-centered is a vulnerable option, as the inhabiter uses feelings to guide its choices, which are merely transformed neurotransmitter data experienced by the inhabiter while thinking and when the central region of its beam of primordial consciousness is illuminating electromagnetic waves at the subtle conscious level. The computer brain continually releases its neurotransmitters and needs these programs to function as a unit. This method of functioning in conditioned existence works well for the normal matter computer brain because it does not think with cognitive states. However, the dark matter inhabiter is a very different energy from normal matter. If the inhabiter uniquely adopts the computer brain's method and uses transformed neurotransmitter data it cognizes as feelings as its guide in making its decisions, then the inhabiter will consistently find itself adrift, lost in the tides of thoughts of the psyche self-centered, experienced as a nonlucid dream or nonlucid nightmare. Thus, the inhabiter can remain an inept participant in the interaction with normal matter by habitually choosing the psyche self-centered. Ergo, the inhabiter will experience the cognitive states of this psyche darkly as a nonlucid nightmare or be deluded into thinking it's in control of its life with a nonlucid dream. All the while, the inhabiter enables its ego-clinging imprint strings through uneducated choices that will cause immediate or the guaranteed further returns of suffering.

The good news, however, is that the living beings that appeared in three-dimensional holograms and created the illusion of relationships between the pawn and "others" represent a magnificent dimension of reality deeply rooted in everlasting bliss, love, and compassion. Each inhabiter is a transformed focal point of the same universe's dark energy substrate layer that has explored its interactions with normal matter in a myriad of complex scenarios facilitated by entanglement. Thereby, the universe's dark energy substrate layer keeps track of interactions between dark matter inhabiters and normal matter computer brains but only as physical attributes,[503] not illusory appearance of individuals or limitations based on color, country, species, or time. Hence, relationships do not mean interactions between distinct, intrinsically existing objects such as living beings but a network of normal matter computer brains and focal points of dark matter that communicate while attached to the two-dimensional grid of the universe's dark energy substrate layer. What empowers the inhabiter is its ability to make favorable choices and reap the reward of accumulating wealth in the form of enabled non-ego-clinging imprint strings and the inhabiter's disinterest in using the vulnerable option of feelings as the means to guide its choices. Instead, the inhabiter uses behavior clues, and as the invisible mental detective behind the scenes of illusory reality, it discerns its motivations. The inhabiter knows which behavior of the pawn in words and actions benefit other living beings. When these behavior clues are detected by the inhabiter, along with the supporting motivation that the inhabiter is disinterested in benefiting the pawn at another living being's expense, then the inhabiter has discerned altruistic behavior clues. Thus, by assembling these behavior clues, the inhabiter is acting as a top-notch mental detective. In addition, the inhabiter's existence will consist of more than just a series of mental functions while embodied within a normal matter computer brain. Instead, the inhabiter is choosing to experience its life as a lucid dream of conditioned existence. When the inhabiter selects the psyche altruistic, then its primordial reactions are tethered to its cognitive reactions in a way that creates mental stability and sustained happiness. This is the first reward commonly experienced by the inhabiter, but with consistent favorable choices, the inhabiter will profoundly bend its perception of itself, the concept of "self" and the pawn.

The living beings, nature scenes, sunrises, and sunsets conjured up at the subtle conscious level of the layered electromagnetic radiation are apparitions without any reality. Knowing this, the inhabiter does not get attached to them but can still relate to them and enjoy their presence[420] and can do so without any notion of I or self.[421] Thus, through choice, the inhabiter gains the ability to have a lucid dream while interacting with normal matter. Motivation is not an illusion but is simply a catalyst, and the inhabiter uses it when choosing a psyche. When the inhabiter prefers not to interact with normal matter in a nonlucid dream or nonlucid nightmare while wiggling in the quicksand, struggling with the cognitive states of the psyche self-centered, the inhabiter opts for the alternative of the psyche altruistic. Having made this choice, the inhabiter's perspective of the illusory world is completely turned upside down, and the inhabiter possesses the aptitude that allows it to deal with its maladies of perception.[422]

When the pawn is correctly understood to be the avatar in its direct interaction with normal matter and not the inhabiter, the inhabiter will be motivated to search for the behavioral clues where the pawn treats all characters, regardless of size, race, color, or species, with love and compassion. Of course, this includes what the immersed inhabiter, having chosen the psyche self-centered, previously believed to be enemies and insignificant others. In fact, these characters in particular are the inhabiter's brightest candles because of what they reveal about the inhabiter's progress in reemergence and its ability to make alternative choices. When the inhabiter is able to put the pawn's needs last with respect to other characters, the inhabiter puts its priority of balancing the equation with normal matter first. When the inhabiter selects fewer creature comforts for the pawn, the inhabiter has transformed its habitual modus operandi of choosing the psyche self-centered. With this transformation, the inhabiter regards its participation in the game of life as its responsibility, while competing against itself and past performance by making alternative choices. Though meditative training may initially be felt by the inhabiter as being difficult, feelings are no longer the determinant of the inhabiter's decisions, because the inhabiter is having the lucid dream of the psyche altruistic. Therefore, the inhabiter does not cowardly refrain from the work but instead shoulders it by training itself so as to gain the wisdom that dispels illusions. Though there are no actual superheroes that act as guardians of the galaxies except within the psyches, there definitely exist courageous inhabiters. A courageous inhabiter is one that addresses its maladies of perception by choosing the psyche altruistic and is willing to experience this lucid dream as a lucid therapeutic nightmare. Because to conquer the sources of suffering while embodied within a normal matter computer brain, the inhabiter cannot actually run or hide. This is where accurate perception is of benefit, because the inhabiter does not need to run or hide from anything it fears or be victimized by its own ignorance that causes suffering. The inhabiter has everything it needs as part of its configuration to fight its own fear and ignorance. By using its beam of primordial consciousness with a central region to illuminate the darkness that harbors symbolic skeletons in closets or a malicious clown that passes through storm drains[423] of darkened cognitive states, the inhabiter exposes these nightmares for what they are. However, it requires the inhabiter's zeal to step into the darkness, though not alone or having to be afraid. The inhabiter's real source of power is within its configuration and is its focal point of dark energy. Thereby, a primordial presence is always available, such as the breath, that creates the inhabiter's confidence and allows the inhabiter to witness any problematic sense impression or mental event by recreating them within the psyche altruistic, lucid therapeutic nightmare.

By consistently choosing the psyche altruistic, the inhabiter will understand what creates unconditional happiness while experiencing its life of conditioned existence as a lucid dream. This leads to the inhabiter's liberation via a choice of enlightened cognitive states. More importantly, when the inhabiter is able to complete its odyssey of discovery but chooses not to and instead keeps one ego-clinging imprint string enabled, the inhabiter transforms itself into a real guardian of the galaxy. Remaining within conditioned existence for however long is necessary to be of benefit as an enlightened inhabiter. When an enlightened inhabiter's journey is complete, a single ego-clinging imprint string is disabled, and non-ego-clinging imprint strings get absorbed by its focal point of dark energy. Devoid of imprint strings, there is nothing demarcating the peripheral surface of an attached focal point of dark energy from the universe's dark energy substrate layer. Therefore, the dimensions of a focal point are smoothed away and become a part of the enduring universe's dark energy substrate layer that is the orchestrator of the innumerable games of life. This facilitates the communication between inhabiters by the phenomena of entanglement, thereby forwarding the transformation of the universe's energy composition by a focal point of change, such that there is more dark and normal energy and less dark and normal matter. This accelerates the expansion of pure consciousness in the universe, as the consciousness of dark energy is a property within itself and is not diluted as the inhabiters complete their journeys.[424] Instead, this act is a compassionate force that expands the invisible universe and accelerates the disappearance of enabled imprint strings, created conscious states, and normal matter computer brains.

In the real universe, there are no pawns or living beings surrounded by phenomena of solidity, and it is without rules, controls, borders, and boundaries but is a place where anything is possible.[425] When a focal point of this universe's dark energy becomes a part of an inhabiter's configuration and the inhabiter consistently chooses the psyche altruistic, then the inhabiter controls its life experience. This choice creates a world with the cognitive states of this psyche where there is no end to love, compassion, happiness, and life itself, because the inhabiter knows it's not the pawn. This is what makes the dark energy that resides and fills this universe so magnificently powerful and why the life experienced by each dark matter inhabiter embodied within normal matter computer brains is important. The inhabiter's arsenal for preventing an uncertain future is not guns, slander, anger, or hate. What can help an inhabiter alleviate its suffering[426] is total freedom from the grip of disruptive emotions, with an inner joy and understanding of the world of phenomena.[427] Every moment represents an inhabiter's opportunity to balance or unbalance an equation while interacting with normal matter; however, in a nonseparable universe, there is more than just one equation. Rather, there are interactions of innumerable proportions. If the inhabiter uses these moments to disable its ego-clinging imprint strings, then the inhabiter's life of conditioned existence will be included as a heroic poem, added to the long story being written by each inhabiter while interacting with normal matter. The inhabiter's contribution and poem will be characterized by highly developed detective skills and with rhythm that expresses the inhabiter's imaginative altruistic means by which it benefited the nonseparable universe. This is life worth living and an equation worth balancing, and the inhabiter that selects this lucid dream of the psyche altruistic will not regret it. This is an accurate description of reality.

No end … only transformations …

Key Point

- To all inhabiters: May your conscious states be those of the psyche altruistic and your life of conditioned existence be experienced as a lucid dream of abundant love, unrestricted compassion, and profound transformation.

During the nonmanifest state, the inhabiter will perceive very bright light, like rolls of brocade being unrolled in five intense colored lights—white, red, blue, green, and yellow. This light is the radiance of the intrinsic nature of the inhabiter's primordial consciousness and the luminosity aspect of the inhabiter's focal point of dark energy.[428]

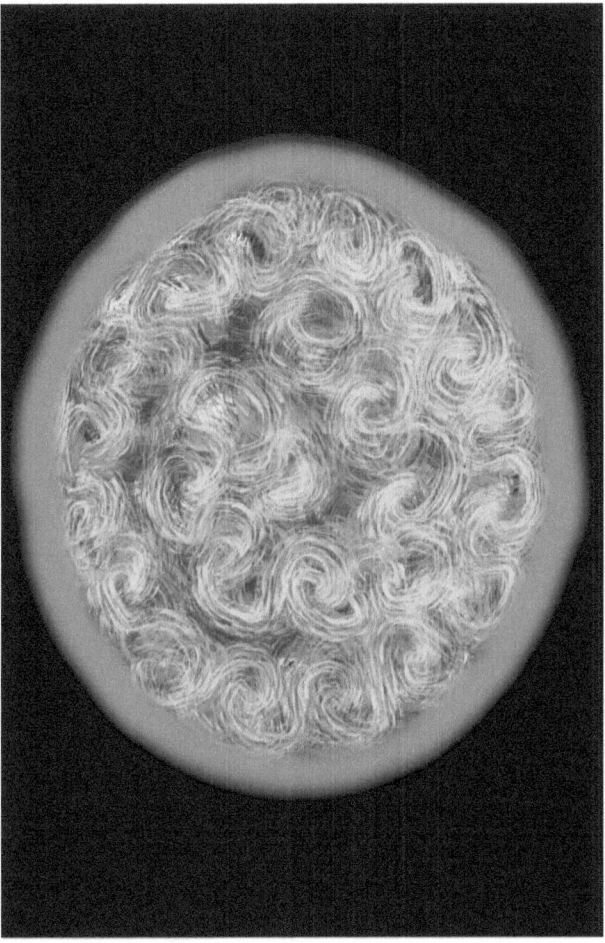

The radiance of the intrinsic nature of the inhabiter's primordial consciousness.

APPENDIX A: QUANTUM TRANSFORMATION REMINDERS

Feelings

The computer brain is able to produce and process data while functioning as a unit because of its neurotransmitters, which are communicative programs that allow neurons to transfer information as a nerve impulse to another nerve fiber, a normal matter muscle fiber, and other normal matter structures. Enabled imprint strings will transform the computer brain's information, which includes details about the neurotransmitters involved in the creation of particular data into electromagnetic waves. Transformed neurotransmitter data act as a signal within electromagnetic waves that draws the attention of the inhabiter's focal point of dark energy with primordial consciousness and determines its reaction, which is to illuminate electromagnetic waves. When the electromagnetic radiation of both psyches is layered, the inhabiter's beam of primordial consciousness will illuminate electromagnetic waves of both psyches up to the subtle conscious level. Whereas, when transformed neurotransmitter data is detected by the central region of the inhabiter's beam of primordial consciousness, the electromagnetic waves created during one cognitive cycle (two rotations of the four steps) will be illuminated for one psyche at a time where associated processing leads. Thus, neurotransmitters are vital to the computer brain's processing and are equally as important to the inhabiter because when neurotransmitters are transformed, they serve as a mental bond that tethers the inhabiter's primordial reaction to the inhabiter's reaction while thinking with created cognitive states.

The inhabiter's awareness of the psyches begins as a feeling that accompanies the making of any kind of image—visual, auditory, tactile, or visceral.[1] While thinking, the inhabiter comprehends transformed neurotransmitter data as feelings of pleasant, unpleasant, and neutral. From a practical sense, the inhabiter makes no distinction between its awareness of subsequent cognized processing as a physical sensation versus a mental event;[2] however, while the inhabiter thinks, it will do so with a feeling that precedes a sense impression or mental event. The inhabiter apprehends physical sensations, experiences emotions, impulses, and urges, and has thoughts and memories because of transformed neurotransmitter data that the inhabiter cognizes as feelings (feeling tones). Although on the subtle conscious level of the layered electromagnetic radiation objects emerge as three-dimensional holograms, the inhabiter's attachment and aversion arise in its reaction to the feelings of pleasant and unpleasant rather than to the objects displayed within the psyche self-centered. What distinguishes the psyche self-centered experienced by the inhabiter as a nonlucid dream from that of a nonlucid nightmare is the degree by which the inhabiter unbalances the equation of interaction by its choice of this psyche. When the inhabiter experiences the psyche self-centered as a nonlucid nightmare, this means that the inhabiter is struggling while thinking with its cognitive states, and this will interfere with the inhabiter's ability to interact with the normal matter computer brain beneficially. Thereby, the interaction between dark matter and normal matter becomes problematic (life interference), but only the inhabiter actually cognitively suffers a nonlucid nightmare of the psyche self-centered.

Though inhabiters suffer in many different ways while thinking with the cognitive states of the psyche self-centered, they are commonly rooted in the choice of this psyche by which to experience momentary reality. There are times when inhabiters make the choice of the psyche self-centered, and they become cognitively anchored to either a physical sensation or a mental event that becomes problematic. Thereby, inhabiters incorrectly assume that sensorimotor information, thoughts, memory, and emotion precede their feelings, which is not the case. Therefore, an incorrect assumption becomes an inhabiter's erroneous belief that produces its delusion that living beings, places, and things displayed as three-dimensional holograms are the cause of pleasant or unpleasant feelings. Ergo, inhabiters are motivated to choose the psyche self-centered repeatedly because they are unable to see past the choices they do not understand. Compounding inhabiters' interpretive dilemma is the fact that the computer brain uses particular neurotransmitters when processing similar information. Correspondingly, an inhabiter, while thinking with cognitive states, will experience similar unpleasant or pleasant feelings associated with prior sense impressions or mental events when these repeatedly emerge within the psyche self-centered. This results in a proliferation of an inhabiter's suffering because of its beliefs that were formulated with a cognitive error. The inhabiter cognitively connects sense impressions or mental events as coming before and to be the cause of its pleasant or unpleasant feelings. By habitually choosing the psyche self-centered, the inhabiter perpetuates the cognitive problematic that does not disappear; instead, the themes that cause the inhabiter's struggle while thinking and that interfere with its life of conditioned existence remain stable. Nonetheless, from the inhabiter's misguided viewpoint, its continued struggle only confirms what it has misinterpreted. Hence, what the inhabiter thought was a viable observation is the inhabiter's incorrect appraisal of a problematic issue. The inhabiter identifies the pawn as being "self," and an escalation of difficulty ensues to produce what the inhabiter experiences as an overimportance of its physical sensations, thoughts, memories, and emotions. The inhabiter suffers because it believes that a feeling (unpleasant or pleasant) with a physical sensation, thought, memory, or emotion means something about "self" and warrants further action/processing that the inhabiter self-actualizes by choosing the psyche self-centered.

To understand the dimension of how problems might manifest while an inhabiter thinks with the cognitive states of the psyche self-centered, use the chart below. You are an inhabiter and to use the chart, start by recreating from your own experience a cognitively problematic event. Systematically break it apart via the chart while keeping in mind that the causal feeling (cognized transformed neurotransmitter data) always precedes a sense impression or mental event. The subsequent four columns represent the resulting effects that can be arranged in different order and depends on the inhabiter. Start the process by asking yourself if your problem included a pleasant or unpleasant feeling and write your answer as one word. Then proceed to the next four columns according to the order by which they emerged within the psyche self-centered until you have answered each column. See the example chart as a guide and then use the complete chart to explore your own cognitive "trip up."

Psyche self-centered experienced by the inhabiter as a nonlucid dream or a nonlucid nightmare.

Cause		Effect		
Feeling (cognized transformed neurotransmitter data)	The focus: Common focus of a thought/ memory/emotion	Motivation	Words and actions of the *Pawn* in the "Psyche Self Centered"	The Inhabiter's delusion
Pleasant	Substances (drugs/alcohol)	Attachment-To procure a pleasant feeling that precedes a physical sensation and/or mental event.	Pawn pursues focus.	Misapprehends the cause of the pleasant feeling.

Example: substance abuse.

The complete chart of the psyche self-centered experienced by you, the dark matter inhabiter, as a nonlucid dream or a nonlucid nightmare. Pick from the columns below.

Cause		Effect		
Feeling (cognized transformed neurotransmitter data)	The focus: Common focus of a thought/ memory/emotion	Motivation	Words and actions of the *Pawn* in the "Psyche Self Centered"	The Inhabiter's delusion
Pleasant	- Living being - Inanimate object - Concept of self - Sex - Money - Food - Substances (drugs/alcohol) - Religious interest	Attachment-To procure a pleasant feeling that precedes a physical sensation and/or mental event. Aversion-To avert a pleasant feeling that precedes a physical sensation and/or mental event.	Pawn pursues focus. Pawn avoids focus.	Misapprehends the cause of the pleasant feeling.

Pleasant feelings.

Cause	Effect			
Feeling (cognized transformed neurotransmitter data)	The focus: Common focus of a thought/ memory/emotion	Motivation	Words and actions of the *Pawn* in the "Psyche Self Centered"	The Inhabiter's delusion
Unpleasant	- Living being - Inanimate object - Concept of self - Sex - Money - Food - Substances (drugs/alcohol) - Religious interest	Attachment-To procure an unpleasant feeling that precedes a physical sensation and/or mental event. Aversion-To avert an unpleasant feeling that precedes a physical sensation and/or mental event.	Pawn pursues focus. Pawn avoids focus.	Misapprehends the cause of the unpleasant feeling.

Unpleasant feelings.

Now it's up to you, inhabiter, to make the choice of the psyche altruistic, which will be easier to do because you understand the illusion of the game of life that appears in three-dimensional holograms at the subtle conscious level of the layered electromagnetic radiation. Many problematic events are effectively addressed simply by you choosing the psyche altruistic and experiencing it as a lucid dream. However, for cognitive events of the psyche self-centered that are problematic because they involve intrusive physical sensations or thoughts you do not like experiencing, which may include impulses and urges, then choosing the psyche altruistic and experiencing it as a lucid therapeutic nightmare (exposure) might be the best course of action to address this problem (discussed in chapter 10, "The Primordial Beam and the Dream or Nightmare").

Keep in mind that in order to overcome your maladies of perception with the psyche altruistic, enabled ego-clinging imprint strings must be disabled. When you choose one of the psyches to experience momentary reality, the cognitive states of that psyche define situations as real. As your choices enable and disable your imprint strings that send signals to the normal matter computer brain and result in alternative programming or allow the computer brain to process data according to its nature, the situations perceived by you are real in their consequence. This means take some time to think through your course of action to address your malady of perception before executing your plan via the psyche altruistic.

Additionally, do not be misled by a common glitch in the interaction with normal matter: there may be times where an inhabiter experiences an unpleasant feeling that precedes a negative mental event (emotion/thought/ memory). This happens and should not be confused with the fact that the preceding feeling in actuality is cognized transformed neurotransmitter data and is not the mental event. Rather, a feeling is a separate phenomenon that occurs before the mental event. Finally, to ensure the execution of your plan is highly effective, you will need a reminder of what you are, which is not the pawn. Although you might not be at the stage in reemergence where you can accept this fact, make sure to include a reminder such as the breath in the periphery of your awareness each time you are perceiving a mental event in the cognitive states of the psyche altruistic. An inhabiter might think of the sensation of the breath as the background music while perceiving the cognitive states of the psyche altruistic at the subtle conscious level of the layered electromagnetic radiation.

The fact of the matter is that neurotransmitters are produced every time the computer brain's circuitry is activated, and transformed neurotransmitter data will be incorporated into every electromagnetic wave created by the inhabiter's enabled imprint strings. Thus the inhabiter will have experienced a preceding feeling every time it is aware of a sense impression or mental event. Although the inhabiter's primordial reactions are dictated by transformed neurotransmitter data, the inhabiter's primordial reaction is not always the same as it relates to the beam, central region, and whether the transformed neurotransmitter data is detected by the focal point of dark energy in the electromagnetic waves of the psyche self-centered or the psyche altruistic.

In order to participate in the game of life while interacting with the normal matter computer brain, the inhabiter must think with its created cognitive states. To know which psyche it is perceiving before making a selection, the inhabiter cannot block feelings, physical sensations, thoughts, memories, or emotions. Although it is true that the inhabiter will not know from moment to moment what might emerge in its created cognitive states, the inhabiter has a choice by which psyche to experience its momentary reality. A wise inhabiter willingly experiences feelings that precede a sense impression or mental event but does so without further subsequent cognitive processing, thus not selecting the psyche self-centered. Accordingly, where there is no attachment or aversion due to a cognitive error, the inhabiter does not feel driven to make a nonbeneficial selection that results in mental proliferation; thereby, causal suffering is not experienced.[3]

The inhabiter benefits when it stops making faulty correlations because it knows feelings come before a sense impression or mental event. Ergo, the sense impression and mental event do not cause the feeling, but without the feeling, a physical sensation, thought, memory, emotion, impulse, or urge would not emerge within the psyche. Comprehending this fact, the inhabiter does not misinterpret its interaction with normal matter and is aware that the computer brain uses particular neurotransmitters when processing similar information. Therefore, the inhabiter's alternative response is due to its acceptance of feelings for what they are (cognized transformed neurotransmitter data), and the inhabiter simply acknowledges that a sense impression or mental event follows the feeling and has occurred. Needing no further processing, the inhabiter does not catastrophize thoughts or use feelings to guide its choices; rather, the inhabiter opts for a rational response of the psyche altruistic. There is no doubt that the inhabiter's choice is rational due to the normal matter computer brain's calculation for impermanence. There are skills that the inhabiter needs to learn and practice before this critical event. The means by which to accomplish these tasks are within the cognitive states of the psyche altruistic.

The Deeper Dimension of "Others"

What is known is that there are many inhabiters embodied within normal matter computer brains spread throughout the universe, and they communicate via the phenomena of entanglement that is orchestrated by the universe's dark energy substrate layer. When inhabiters choose a psyche by which to experience reality the electromagnetic radiation of both psyches dissolves into their focal point of dark energy and produces a reverberation that disables and enables their imprint strings. After their focal point of dark energy has acquired the electromagnetic radiation, sometimes information about the pawn within the psyche chosen by the inhabiter is transformed to a quantum signal. This signal is then disseminated by the universe's dark energy substrate layer to other inhabiters and incorporated into both psyches in a way that contextually makes sense to the inhabiter. Because an inhabiter's cognitive states are created at an extremely rapid rate, there is only an imperceptible instant by which the inhabiter actually perceives the signal received by the universe's dark energy substrate layer that has not been adulterated by the normal matter computer brain. Once the normal matter computer brain receives signals from the inhabiter's enabled imprint strings, it transforms it to neural synchrony, modifies the information, and maps it with its circuitry. The normal matter computer brain will then use the information while processing for the core-I character and autobiographical-I character. This means the computer brain adds a lot of information to the signals it received from the inhabiter's enabled imprint strings.

The inhabiter can surmise that "other" living beings perceived in the psyche as directly interacting with the pawn came about from quantum signals it received from the universe's dark energy substrate layer. Thereby, at some point, the inhabiter has been communicating with other inhabiters spread throughout the universe. When the inhabiter perceives in the psyche the pawn listening or watching media where other living beings have never been directly encountered by the pawn but rather are heard or seen via a computer, radio, television, or on a movie screen, the inhabiter could accurately conclude that these characters did not come about from quantum signals received from the universe's dark energy substrate layer. Instead, these characters came about from the network of normal matter computer brains that are exchanging information they received from the inhabiter's enabled imprint strings. Whichever the case might be, the inhabiter must also understand that other living beings and the situations involving these characters have no actual true features. Accordingly, the normal matter computer brain has modified the universe's dark energy signal and the signals it has received from the network of computer brains in a way that is unknown to the inhabiter. Be that as it may, the inhabiter can still make use of the experience while witnessing other characters with the pawn in the psyche altruistic. Without reifying or making assumptions about what cannot be known, through other living beings the inhabiter can contemplate what a round of the game of life might be like with a pawn of similar and unrelated species. The inhabiter is able to garner information when the living being is a human and shares its life experience and its actions are perceived by the inhabiter. When the living being is not human, the inhabiter can pick up clues by the character's behavior and situations that the character is in. However, in both of these scenarios, the inhabiter is picking up these as indirect clues because the inhabiter only directly experiences sensorimotor information (smell, tactile sensation, sound, sight, and taste) through the pawn's body and not through the bodies of these other living beings.

Thus, deeper dimension of others is accurately perceived by the inhabiter through the cognitive states of the psyche altruistic, which allow an inhabiter to utilize imagination while simultaneously taking into consideration the limitations of this imagined mental event. Hence, it is not a replication of sensorimotor information or the inhabiter's direct experience that grants the inhabiter the gift of profound insight. Instead, through the three-dimensional holograms of other living beings at the subtle conscious level, empathy is the catalyst for the inhabiter's transformative experience. With wisdom, a dark matter inhabiter understands that the primordial consciousness of its focal point of dark energy and its enabled imprint strings utilizing the normal matter computer brain's data created these other living beings. Nevertheless, these characters that emerge within the psyches are there for an important reason. Because when the inhabiter chooses the psyche altruistic, it can imagine the suffering experienced by other parts of itself, as dark matter inhabiters, lost within the vast universe in nonlucid dreams and nonlucid nightmares of the psyche self-centered. Compassion that comes about from identifying with the plight of others and being relieved from the delusion of self is what motivates the inhabiter to continue to choose the psyche altruistic, so as to benefit the nonseparable universe.

Four Reliable Sense Impressions

There are four sense impressions that the inhabiter can reliably identify both at the gross conscious and subtle conscious level of the layered electromagnetic radiation. These are movement, solidity, temperature, and moisture. The inhabiter might choose to think of these as its nonconceptual (gross conscious) and conceptual (subtle conscious) sense of the four fundamental forces.

Movement (electromagnetism): The inhabiter may experience this as a nonconceptual or conceptual flicker in its visual field. This effect is created by the inhabiter's beam of primordial consciousness detecting transformed neurotransmitter data and reacting by generally illuminating electromagnetic waves while the central region of the beam reacts to the transformed neurotransmitter data by specifically illuminating electromagnetic waves. When thinking of this in terms of the fundamental forces, movement would be replaced by the term electromagnetism.

Solidity (gravity): The inhabiter is nonconceptually or conceptually aware of the sensation of being pulled down as if it is being anchored in solidity. The inhabiter is perceiving the effect of the layered electromagnetic radiation held in place by the kinetic energy produced by its enabled imprint strings. Thus, instead of floating away, the layered electromagnetic radiation levitates above the surface of the inhabiter's focal point of dark energy. When thinking of this in terms of the fundamental forces, solidity would be replaced by the term gravity.

Temperature (strong force): The inhabiter is nonconceptually or conceptually aware of differences in temperature. When thinking of this in terms of the fundamental forces, temperature represents the strong force within the electromagnetic waves of the layered electromagnetic radiation. The strong force is created by the imprint strings or magnetic monopole that is the most enabled.

Moisture (weak force): The inhabiter is nonconceptually or conceptually aware of the sensation of moisture. When thinking of this in terms of the fundamental forces, moisture represents the weak force within the electromagnetic waves of the layered electromagnetic radiation. The weak force is created by the imprint strings or magnetic monopole that is the least enabled.

- Temperature as the strong force and moisture as the weak force can certainly be cognitively switched. Whichever sense impression is the greatest felt by the inhabiter would be the strong force.

Correspondingly, the sense impression that is weakly felt by the inhabiter would be the weak force.

- If the inhabiter becomes aware that it is once again thinking of itself as being the pawn, it regroups and widens its viewpoint with the four reliable sense impressions.

The Mental Detective

Reemergence is a process by which the inhabiter liberates itself from being immersed in the cognitive states of the psyche self-centered via its response and choice of the psyche altruistic. When just beginning its process of reemergence, the inhabiter, still thinking of itself as the pawn, asks itself what the metaphorical role model would do in a particular situation, and this serves to guide the inhabiter to choose the psyche altruistic. However, as more of its non-ego-clinging imprint strings are enabled and ego-clinging imprint strings are disabled, the idea of self and being the pawn is abandoned by the inhabiter. An inhabiter with an accurate relative understanding that it is a focal point of dark matter embodied by a normal matter computer brain has a simplified picture of what and where it is and uses this to strategize balancing the equation of interaction with normal matter. Furthermore, this inhabiter is aware of its beam of primordial consciousness with a central region that illuminates electromagnetic waves of the psyches sequentially. With the background music of the breath in the periphery of its awareness, the inhabiter becomes aware of a feeling tone of pleasant, unpleasant, or neutral and simply acknowledges this confirms the computer brain's programs are loaded and running due to the work performed by its enabled imprint strings. Next, the inhabiter evaluates what it is experiencing and witnessing as a subsequent sense impression or mental event. The inhabiter reminds itself that the preceding feeling and subsequent sense impression or mental event has emerged within the psyche self-centered and the psyche altruistic because electromagnetic waves with incorporated transformed neurotransmitter data are illuminated by its beam of primordial consciousness with a central region.

Though reality is symbolically displayed at the subtle conscious level, if the inhabiter perceives clues indicating cruelty, corruptness, hatred, jealousy, and self-indulgence, this alerts the inhabiter that the computer brain is directing itself and its enabled ego-clinging imprint strings are creating an interpretation of the data. The inhabiter does not fault the computer brain that unknowingly and according to its nature is directing itself, nor is the inhabiter upset by its enabled ego-clinging imprint strings, as they are fulfilling a purpose. However, selecting the psyche self-centered will not balance the equation of interaction with normal matter; rather, the computer brain will continue to direct itself, and the inhabiter remains immersed in nonbeneficial cognitive states. Therefore, without trying to block what it perceives, the inhabiter instead calmly lets three-dimensional holograms of the psyche self-centered arise and pass away while patiently waiting for the alternative display of the psyche altruistic. Altruism can be portrayed in many different ways via the pawn and other living beings. When benevolence, humanitarianism, kindness, and selflessness are perceived by the inhabiter, this alerts it that the computer brain's data has been altered by the inhabiter's primordial consciousness, and non-ego-clinging imprint strings are creating an interpretation of the data. The inhabiter that responds by selecting the psyche altruistic is transforming its life of conditioned existence via altering the computer brain's programming with its enabled non-ego-clinging imprint strings.

Accordingly, there are relatively straightforward clues that the inhabiter as the mental detective can pick up before making a psyche selection, whereas other times it might not be so obvious. There are scenes of the psyche altruistic at the subtle conscious level meant to give the inhabiter an opportunity to respond differently and disable rather than enable its ego-clinging imprint strings. Thereby, the inhabiter will be exposed to a training session designed as a recreation within the psyche altruistic that includes unpleasant feelings that precede three-dimensional holograms of challenging mental events the inhabiter previously mistook to be obstacles and enemies. In order to not respond in a similar fashion, making a nonsensical selection, the inhabiter will need to polish its skills and use the mental clue of motivation when challenged with a psyche altruistic recreation. The dark matter inhabiter must always keep in mind that it is a very different energy than normal matter and that the computer brain does not comprehend its own existence let alone the inhabiter's.

Two Different Energies Interacting

Dark Matter Inhabiter	Normal Matter Computer Brain
Beam of primordial consciousness with a central region	Claustra's meta-algorithm
Focal point of dark energy	Circuitry made of normal matter.
Enabled imprint strings	Neurons
Transformed neurotransmitter data/Feelings	Neurotransmitters

Comparing the dark matter inhabiter to the normal matter computer brain.

When the inhabiter is motivated to treat normal matter through its choices generously, affectionately, and sympathetically, the inhabiter is giving the computer brain a symbolic voice while interacting. After all, normal energy was free in the universe before it was drawn by an inhabiter's enabled imprint strings and transformed to a normal matter computer brain. To engender a motivation that balances the equation of interaction through the beneficial choice of the psyche altruistic, the inhabiter reminds itself that every animate or inanimate object it perceives is made from the transformed data of a normal matter computer brain or from the network of normal matter computer brains.

The conditioned existence of an inhabiter embodied within a normal matter computer brain is life with an extremely high level of complexity. The electromagnetic waves with incorporated transformed neurotransmitters created by enabled imprint strings contain the computer brain's data that, when illuminated by the inhabiter with its beam of primordial consciousness with a central region, allows the inhabiter to experience conditioned existence as if it's a living being. When the normal matter computer brain is configured to produce data for a human psyche, the inhabiter experiences life as a human even though it is really a focal point of dark matter and only derives a sense of self because of what it feels through the pawn's body. Without the normal matter computer brain, the inhabiter would not think with the cognitive states of a psyche and would not believe itself to be alive and existing as a human. Without the inhabiter, the normal matter computer brain would not exist. Both the normal matter computer brain and the dark matter inhabiter react to their life of conditioned existence but in their own unique way. The universe's dark energy substrate layer does not think with created cognitive states, yet when morphed and transformed to a dark matter inhabiter, it does. Accordingly, one might deduce that the dark matter inhabiter thinks with artificial intelligence. The normal matter computer brain participates in its life of conditioned existence but not cognitively, rather by reacting to degrees of neural synchrony. Because the computer brain does not know it or the inhabiter exists, one might believe that normal matter is mimicking being alive. However, there is an alternative viewpoint that takes into consideration that the normal matter computer brain reacts and therefore it can be construed as being alive, as there is obviously more than one way to live in the universe.

Thus dark matter and normal matter are two energies that while interacting are very intertwined. Signals received by the computer brain from enabled imprint strings become a part of normal matter's large repertoire of computations. When the computer brain's signals are received by the inhabiter's enabled imprint strings, normal matter's computations are seamlessly handed off to the inhabiter, with only feelings to announce to the inhabiter that this has happened. Thereby, was deliberate opaqueness[4] one of the variables designed into the equation of interaction by the universe's dark energy substrate layer? Perhaps the synergy between dark matter and normal matter has evolved to this level of complexity, or alternatively it has always been present and serves as a measurement of an inhabiter's understanding, whereby opaqueness is transformed to clarity. If the inhabiter is to understand the complexities surrounding its life of conditioned existence, its relative and infinite viewpoints must be connected. It is the concept of self and the associative possessive belief of an inhabiter that concludes erroneously that it has exclusive rights to illusory phenomena within the psyche self-centered. Which energy, dark matter, or normal matter actually owns either of the ungraspable psyches, where each of the two energies contribute to the emergence of the *psyche phenomena* that displays three-dimensional holograms at the subtle conscious level? What matters is that when an inhabiter believes itself to be the pawn and human, sometimes being human becomes a measuring stick by which the inhabiter values life while thinking egocentrically. Though it is true the inhabiter is a bit human since it thinks with transformed data for a human psyche, the inhabiter through its individualistic viewpoint is transforming itself to be less like dark energy. The dark matter inhabiter's primary attribute and potential benefit to the normal matter computer brain is that it is morphed dark energy with primordial consciousness. The accessory in the game of life that allows the inhabiter to participate in the game and directly alter the interaction with normal matter so as to balance the equation is the pawn. Accordingly, a human pawn implies that the normal matter computer brain's configuration is intricate. Thereby, the normal matter computer brain is able to handle complex calculations needed to support the dark matter inhabiter with primordial consciousness that is transforming itself into a skillful mental detective so as to disable its ego-clinging imprint strings. This is the crux of the importance of the pawn being human, which is the avatar of the inhabiter's and the normal matter computer brain's direct interaction, but where only the inhabiter is the energy that can train itself to use the pawn for benefit. In order for the inhabiter to truly benefit itself and normal matter, the inhabiter must feel empathy for the normal matter computer brain. Correspondingly, the inhabiter must transform itself from energy that knows it is alive, not only because it senses this perception through the pawn's body but because it is aware of its existence without a subject/object reference. The inhabiter gains this skill through the cognitive states of the psyche altruistic.

The dark matter inhabiter's relationship with normal matter is tenuous because, though the computer brain is unaware of the inhabiter's presence, it can create the data that can push the inhabiter's "immersive buttons." Part of all inhabiters' path to reemergence is grappling with the fact that while immersed in the psyche self-centered, they unbalanced the equation of interaction and not just as an individual. As there are many focal points of dark matter spread throughout the universe as inhabiters lost in nonlucid dreams and nonlucid nightmares of the psyche self-centered accordingly, an immersed inhabiter through its choices and on a microscopic level has essentially weaponized cognitive states by selecting the psyche self-centered.[5] Thereby, that inhabiter unbalances the equation of interaction with a normal matter computer brain. Although immersion is an unavoidable consequence of an inhabiter living a life of conditioned existence, the duration of the immersive experience is dictated by an inhabiter's choices. The inhabiter as part of the population of focal points of dark matter can extrapolate its microscopic weaponization of cognitive states to that which has the potential to marshal negative effects at the macroscopic level. The inhabiter does this by hindering another inhabiter's ability to balance their equation of interaction with a normal matter computer brain somewhere in the nonseparable universe.

The perspective of an inhabiter training itself to be a mental detective and orchestrating its reemergence might be portrayed as an energy with lucid awareness such as this: the inhabiter is cognizant of its presence in the form of background music, such as the breath, in its peripheral awareness and understands that whatever it feels or experiences is simply a display of its synergy with the normal matter computer brain. The pawn is the avatar that allows the inhabiter to directly experience its interaction with the normal matter computer brain and other living beings, places, and things the inhabiter indirectly experiences in its interaction with normal matter. The dark matter inhabiter committed to balancing the equation of interaction with the computer brain might remind itself of its mission via a word of honor that will not be heard by normal matter.

Scenario one: The pawn is perceived by the inhabiter as being human and alone without another living being present, either human or nonhuman.

The inhabiter's silent pledge to the normal matter computer brain while perceiving illusory phenomenon displayed at the subtle conscious level of the layered electromagnetic radiation: "I, the dark matter inhabiter, perceive our avatar of direct interaction (pawn) surrounded by three-dimensional holograms of illusory places and things but am aware that you, normal matter computer brain, do not. Albeit, I am looking for clues where I perceive our pawn treating illusory phenomenon respectfully, not wastefully and not greedily consuming or using more than it needs to maintain a healthy existence, and does so without purposefully harming another living being. When I perceive these clues, I will know that it is the psyche altruistic I am witnessing and will select this psyche by which to experience momentary reality."

Scenario two: The pawn is perceived by the inhabiter as being human and with another living being or beings of either human or nonhuman species.

The inhabiter's silent pledge to the normal matter computer brain while perceiving illusory phenomenon displayed at the subtle conscious level of the layered electromagnetic radiation: "I, the dark matter inhabiter, perceive our avatar of direct interaction (pawn) and 'other' living beings and am aware that you, normal matter computer brain, do not. Albeit, I am looking for clues where I perceive our pawn treating our illusory phenomena respectfully, not with malice, only with loving kindness. Where there is controversy, I will recognize this as an expected misunderstanding between us; after all, we are two energies that perceive neural synchrony differently. Correspondingly, I will be looking for clues that display creative alterations to neural synchrony orchestrated by my focal point of dark energy and non-ego-clinging imprint strings. This would entail the pawn using nonviolent and innovative means to address a difficult situation. Where there are circumstances in which a living being is struggling for survival, I will be looking for clues where the pawn is seen trying to assist the struggling being without seeking praise and preferring obscurity. These are distinguishing characteristics of the pawn of the psyche altruistic. When I perceive these clues, I will know what psyche I am witnessing and will select the psyche altruistic by which to experience momentary reality.

"Thereby, as a dark matter inhabiter and mental detective, I am motivated to make beneficial choices for us, normal matter computer brain, for however long we have left to interact. Through the interaction, I know we will continue to create these illusory manifestations at the subtle conscious level and that from your perspective these are degrees of neural synchrony. However, from my perspective, the three-dimensional holograms we collaborate to create represent opportunities to disable my ego-clinging imprint strings. In spite of the fact that we have different perspectives, the only way I can directly show you loving kindness is through the pawn of the psyche altruistic. Nonetheless, it is not only you, normal matter, that I need to thank, as there are parts of me somewhere in the universe working on our behalf as enlightened inhabiters. To honor these enlightened inhabiters, I will do so symbolically with the pawn and with a unique characteristic that is a mark of my gratitude. This impression can be represented in many different ways to depict my gratitude.

- The pawn of the psyche altruistic might have the metaphoric role model on its shoulder.
- The pawn of the psyche altruistic might have the metaphoric role model above its head.
- The metaphoric role model might be in the pawn's throat every time I perceive it eating food or drinking.
- The pawn might have a heart that emanates a clear light of compassion.

When I perceive these luminous representations, it will remind me of the enlightened inhabiter that I am transforming myself to be."

There are three questions, depending on how an inhabiter answers them while thinking with the cognitive states of the psyche altruistic, that will reveal its progress in reemergence and transformation:

- What am I?
- Why am I here living a life of conditioned existence?
- What is my life purpose?[6]

Suffering: Lack of Awareness While the Inhabiter Struggles to Think with Artificial Intelligence

The inhabiter's perception of the psyche self-centered" involves habitual reactions to pleasant, unpleasant and neutral feelings that come about when the inhabiter cognizes and categorizes transformed neurotransmitter data (TND). These feelings precede sense impressions (physical sensations) and mental events (thoughts, memories and emotions) that are transient and have no actual reality but emerge within the psyches when the inhabiter thinks with artificial intelligence.

The inhabiter immersed is unaware that it is embodied by a normal matter computer brain and is thinking with artificial intelligence. Correspondingly the inhabiter remains oblivious to the fact that its attachment and aversion arise in its reaction to the feeling state itself rather than to the object the inhabiter cognizes into existence.

Sense impressions and/or mental events do not contain nor constitute any lasting separate entity that could be called a self but the inhabiter immersed in the cognitive states of the psyche self-centered does not adopt this viewpoint. The inhabiter instead believes the pawn is self. The inhabiter's lack of awareness of what and where it is and that it is thinking with artificial intelligence leads to suffering.

The process by which the inhabiter thinks with artificial intelligence using the cognitive states of the psyche self-centered:

- The inhabiter with primordial consciousness detects transformed neurotransmitter data (TND) incorporated within electromagnetic waves. The inhabiter then reacts reflexively to the TND using its ability to illuminate electromagnetic waves with natural intelligence and this allows the inhabiter to think with artificial intelligence.
- When beginning to think with artificial intelligence instead of detecting TND, the inhabiter cognizes them as feelings. The inhabiter then reacts to varying degrees impulsively and categorizes the feelings as being pleasant, unpleasant or neutral.
- When the inhabiter cognizes TND as feelings and categorizes them impulsively as pleasant, unpleasant or neutral this means the inhabiter is initiating a sequence of steps that will allow it to think with artificial intelligence. When the inhabiter thinks, the normal matter computer brain's programs will be loaded and running due to steps performed by the inhabiter's enabled ego-clinging imprint strings and primordial consciousness. Consequently in the psyche self-centered illusory sense impressions and mental events will manifest and the dark matter inhabiter will perceive them.
- When the inhabiter cognizes TND and due to an impulse and /or urge of variable intensity categorizes them, if the feeling is pleasant or unpleasant, part of the inhabiter's reaction is to pursue or avoid the feeling with attachment or aversion, respectively and

with similar intensity as the impulse and/or urge. Glutamate is the most abundant excitatory neurotransmitter of the normal matter computer brain and when part of the mix of TND this neurotransmitter contributes to the intensity of the inhabiter's cognized momentary experience. Thereby the inhabiter impulsively categorizes what it is feeling and reacts with attachment or aversion to pleasant or unpleasant feelings in order to perpetuate the pleasant feeling and/or end the unpleasant feeling.

The inhabiter will:

- Cognize a feeling.
- Experience the feeling.
- Impulsively categorize the feeling.
- If the feeling is pleasant the inhabiter will try to pursue it so that it sustains a pleasant feeling.
- If the feeling is unpleasant the inhabiter will try to avoid it so that it ends the unpleasant feeling.

However, this strategy, when used long term, will result in the inhabiter's suffering because while thinking with artificial intelligence using the cognitive states of the psyche self-centered what the inhabiter is attached or aversive to, is TND. The inhabiter will impulsively categorize TND not just as pleasant but also as unpleasant and neutral feelings and does so unpredictably. The inhabiter cannot avoid TND nor the feelings they create because the normal matter computer brain continually releases neurotransmitters.

The psyche self-centered contains electromagnetic waves with incorporated TND created by the inhabiter's ego-clinging imprint strings. When the inhabiter uses its primordial consciousness to detect these TND and reacts by illuminating these electromagnetic waves it will think with artificial intelligence that serves a purpose. The purpose being that the psyche self-centered is to create a display of symbolic representations (three dimensional holograms) at the subtle conscious level of the layered electromagnetic radiation. These holograms of living beings, places and things represent the inhabiter's relative (subjective) interpretation of a normal matter computer brain directing itself. The inhabiter's dilemma is that the normal matter computer brain directs itself in a way that is unknown to the inhabiter but will produce neurotransmitters continously until death.

After cognizing TND as a feeling and categorizing it, the inhabiter then cognizes the normal matter computer brain's data for a sense impression (physical sensation). The preceding cognized TND that the inhabiter categorized as a pleasant, unpleasant or neutral feeling will set the tone by which the inhabiter will perceive the initial sense impression.

- While thinking with artificial intelligence the inhabiter's ability to detect TND with its primordial consciousness and illuminate electromagnetic waves reflexively far exceeds its ability to cognize TND as feelings and categorize them impulsively as pleasant, unpleasant or neutral. This means regardless of how many or what kind of neurotransmitters are released by the computer brain at any given moment, when transformed by enabled imprint strings and incorporated into electromagnetic waves, there is not one processing event that escapes the awareness of the inhabiter using its primordial consciousness. Albeit, the inhabiter while thinking with artificial intelligence will be limited in the amount of TND it can cognize at any given moment. Ergo, the inhabiter while thinking with artificial intelligence will extrapolate the same feeling (cognized TND) and use it to set the tone when further cognizing. The pleasant, unpleasant or neutral feeling that preceded the inhabiter's initial cognizing of a sense impression (physical sensation), will thereby set the tone of perception for the inhabiter's cognizing of an initial mental event (thought, memory or emotion) and subsequent mental events. Be that as it may there exists the possibility that after the initial mental event the inhabiter may cognize different TND and categorize it as a contrasting feeling. Therefore the initial feeling will be replaced by one of the three feelings (pleasant, unpleasant, neutral) and will set the tone of perception for the subsequent mental event.
- Because the inhabiter reacts to cognized TND by categorizing it impulsively, and when it is pleasant or unpleasant becomes attached

or aversive to the feeling that sets the tone by which an initial sense impression and subsequent mental event(s) is perceived, the inhabiter is often fooled by a glitch in its interaction with normal matter. There may be times where an inhabiter experiences an unpleasant feeling that precedes a negative mental event (emotion/thought/memory). When this happens and due to the fact that the inhabiter's cognitive states are created at an exceedingly fast rate, the feeling and the mental event are melded complexly together. This prevents the inhabiter from recognizing that the feeling (cognized TND) actually came before the mental event.

A thought, memory or emotion in reality is data and though it is cognized by the inhabiter it is ungraspable because the inhabiter is unaware of how it is actually created by the normal matter computer brain. Normal matter's data is not solid and though the dark matter inhabiter cognizes it and thereby perceives it symbolically as living beings, places and things normal matter does not "see" it and there is nothing actually solid about the information as it is contained within electromagnetic waves.

When the inhabiter cognizes TND as a feeling and categorizes it as unpleasant this sets the tone for what follows which might include one of the primary emotions: (happiness, sadness, fear, anger, and surprise) or the secondary emotions of (embarrassment, jealousy, guilt and pride). Notwithstanding while thinking with artificial intelligence, the inhabiter is formulating its unique interpretation of data created by a normal matter computer brain that does not experience an unpleasant feeling nor perceive with emotion and thereby did not create, categorize nor extrapolate the unpleasant tone with the same cognitive intent as the inhabiter. However, the inhabiter through its unique interpretation is tempted to choose to think with the artificial intelligence of the psyche self-centered repeatedly. Instead of the inhabiter interpreting the pleasant or unpleasant feeling that is accompanied by attachment or aversion as meaning it is perceiving the psyche self-centered, the Inhabiter selects it. Consequently, the inhabiter has made a choice to think non-lucidly and with cognitive errors.

Because the inhabiter's cognitive states are created at an extremely fast rate and the inhabiter is unable to suss the preceding feeling from that of the subsequent sense impression and/or mental event, the inhabiter erroneously concludes that the feeling is due to a physical sensation and/or thought, memory or emotion which is not the case.

The inhabiter then compounds the problem with its most difficult cognitive error and that is its most strongly held belief that it is the pawn because it feels sensorimotor data through the character's body. When choosing a psyche the inhabiter has selected cogntive states that create the perception of "real" and when the choice is that of the psyche self-centered, the inhabiter assumes incorrectly that the living beings, places and things have an actual true existence which is not the case. Consequently the inhabiter remains unaware of what it is (dark matter), where it is (embodied by a normal matter computer brain) and that it uses normal matter's data to create what it is experiencing.

The inhabiter's cognizing of TND as feelings and categorizing them as pleasant, unpleasant or neutral sets the tone by which the inhabiter will subjectively perceive. There are times, from the inhabiter's view point where the preceding feeling does not cognitively fit with the subsequent data that is cognized as a sense impression and/or mental event. This means that while thinking with the artificial intelligence of the psyche self centered, the inhabiter experiences a cognitive mismatch. If the cogntive mismatch includes a preceding unpleasant feeling along with an impulse and/or urge that is followed by a sense impression (physical sensation) and/or mental event (thought, memory, emotion), the unpleasant feeling sets the subjective perceptual tone. Consequently some immersed inhabiters using the cognitive states of the psyche self-centered for self examination will worry that what they are experinceing means something unpleasant about self. This means sometimes when the perceptual tone is set by an unpleasant feeling and is followed by the inhabiter's cognizing of data that creates disturbing mental events, a negative intrusive thought will arise, a thought that the inhabiter fears the most, or least wants to have. When this happens the inhabiter will experience the psyche self-centered as a non lucid nightmare and part of its subjective perception might be that the self is horrible for thinking thoughts with disturbing content. The inhabiter's lack of awareness that it is thinking with artificial intelligence and that it is actually reacting to cognized TND experienced as an unpleasant feeling, rather than to the illusory sense impression and/or mental event leads to the inhabiter's suffering of a cogntive mismatch. The inhabiter's suffering from a cognitive mismatch is perpetuated by its habitual response. The inhabiter's habitual response is to indulge the unpleasant feeling by choosing the psyche self-centered and therefore the inhabiter will not be thinking lucidly with artificial intelligence. Instead the inhabiter chooses the artificial intelligence that will perpetuate the inhabiter's delusion and/or suffering.

Motivation is the Key to Accurate Understanding

If the inhabiter is unable to discern what motivates it (feelings) to make choices and only uses as clues what it experiences through the pawn's body and the character's words and actions are identified as self, the inhabiter will make erroneous assumptions that preclude it from addressing its problematic thinking.

The choice to think with alternative cognitive states and the artificial intelligence of the psyche altruistic is what allows the inhabiter to learn a simple fact: The inhabiter will never be able to deal with its struggle to think with artificial intelligence by impulsively indulging feelings (cognized TND) and allowing them to guide its choices. To overcome its difficulties the inhabiter must perceive the full illusion of its created cognitive states while consistently choosing the psyche altruistic. The inhabiter has the ability to do this because it is dark matter with primordial consciousness but in order to adopt an alternative viewpoint of what it is, the inhabiter must experience this realization without a subject/object reference.

Only with consistent choices of the psyche altruistic will the inhabiter train itself to see all joys and sorrows as if it were watching a movie and will adopt the alternative viewpoint that allows it to let go of the idea that it has to strive hard to avoid TND it cognizes as unpleasant feelings and subsequent data it cognizes as disturbing sense impressions and/or mental events. This will transform the inhabiter's previous unpredictable and impermanent happiness to happiness that is stable and indestructible.[7]

However, if the inhabiter is not motivated to learn current theories from many areas of science and adopt wisdom that comes from spirituality, but instead narrows clues to those that it sees, smells, touches, tastes and hears through the pawn's body, it will make erroneous assumptions that preclude it from understanding the deeper dimensions of its reality.

228

The dark matter inhabiter with primordial consciousness interacts with a normal matter computer brain that does not perceive meaning but responds to degrees of neural synchrony repetitively, unambiguously and according to the rules of neural functionality within a finite period of time. The inhabiter's enabled imprint strings selectively chose the computer brain's data based on whether they are creating a sense impression or mental event. In this way the normal matter computer brain's data gets broken down into packets of information. When the electromagnetic radiation is layered, the inhabiter's challenge is to use its primordial consciousness to reassemble the data, and in a way that creates the inhabiter's relative (subjective) interpretation of a computer brain directing itself (psyche self-centered) and the computer brain being directed by the inhabiter (psyche altruistic).

The inhabiter utilizing its primordial consciousness is the designer of the psyches and it captures all the variations of neural synchrony during the period when its enabled non-ego-clinging imprint strings tap in and when the layered electromagnetic radiation dissolves into its focal point of dark energy (discussed in chapter 9, "Psyches in the Wind"). The exact method used by the inhabiter when it reassembles the computer brain's data is a mystery, but without the activity of its primordial consciousness the interaction between dark matter and normal matter could not take place. Significantly the inhabiter makes use of its beam of primordial consciousness configured with a central region to detect TND and reflexively illuminate billions of small elements of information contained in an astounding number of electromagnetic waves. This allows the inhabiter to create for itself patterns in the linkage of data and to follow its own step by step instructions in the form of primordial algorithms. These primordial algorithms provide the inhabiter its reproducible general solutions to addressing the complexities of interacting with a normal matter computer brain and determines the way it sorts, ranks, and methodically scans for abstract patterns. Thereby the inhabiter transforms neural synchrony into the artificial intelligence of created cognitive states within the psyches.

When devising the primordial algorithms of the psyche self-centered the inhabiter follows the computer brain's rules based on degrees of neural synchrony and the inhabiter learns by normal matter's example. Nonetheless when the inhabiter masterminds its primordial algorithms of the psyche altruistic it compassionately bends the rules set by the computer brain and does so in a way that is acceptable to normal matter. Each inhabiter spread throughout the vast nonseparable universe uses its primordial consciousness to interact with a normal matter computer brain. The more complexly configured the normal matter computer brain is that embodies an inhabiter, the more complex the algorithms must be, and to invent them requires the inhabiter's primordial creativity.

The inhabiter's perceptual bubble while thinking with the AI of the psyche self-centered and psyche altruistic is an expression of the inherent creativity of its primordial consciousness.[8] Consequently within the psyches what emerges are behavioral patterns of living beings, nature and phenomena and there appears to be in the fabric of the universe intelligent design. Through phenomena that include living beings, nature, and climate, the inhabiter's primordial consciousness guides it in behavior patterns that will either balance or unbalance the equation while interacting with normal matter. However, due to the fact that dark matter and normal matter are very different energies, there are variables to the equation that the inhabiter will not be able to reify or predict.

- Created conscious states are impermanent and the exact moment that a round in the game of life will end is uncertain.
- The inhabiter cannot reify its primordial consciousness with a subject/object reference and thereby cannot quantify it with instrumentation.
- Although while thinking with a psyche the inhabiter will be able to use mathematical equations to describe how the computer brain's neurons collect information and communicate, there is still much lacking in the inhabiter's understanding of normal matter.

What is certain is that in order to liberate itself and normal matter from their interaction, the inhabiter cannot be deceived by its primordial consciousness. A magician is never deceived by the illusions he created himself, realistic though they may be, because he knows they do not actually exist. In the same way the inhabiter cannot be fooled when it perceives the AI of the psyche it creates.[9]

What the inhabiter initially feels and what follows which includes impulses and/or urges, physical sensations, thoughts, memories and emotions, the inhabiter's primordial consciousness creates before the inhabiter perceives it. The information the inhabiter perceives within an AI will be the basis for its decisions, and the inhabiter's responses in choosing to think with one of the psyches means it has made a momentary decision. Because the inhabiter's cogntive states are made at an imperceptible rapid rate, the inhabiter must compulsively make thousands of decisions in any given moment.[10]

Many inhabiters that perceive phenomena believe they are able to do so because they have eyes, yet neither dark matter nor normal matter has eyes. What the inhabiter actually perceives is what it subjectively interprets with the information contained within the perceptual bubble created by its primordial consciousness. Although every living being the inhabiter perceives only exists within an AI of the psyche, it is through relationships that the inhabiter defines its place in the world and receives beneficial guidance or is led astray. Through voices of the enlightened, altruistic, egocentric, lost, or desperate, living beings are created by the inhabiter's primordial consciousness. The living beings the inhabiter chooses to listen to will guide it to further immersion or liberation.

Whether the inhabiter feels pleasant or unpleasant while perceiving the pawn receiving either flattery or insults, the thoughts the inhabiter perceives while thinking with AI are due to the activity of its pure primordial consciousness. A pawn in a hologram that shouts flattery or insults is not heard by the computer brain but its data that is contained within electromagnetic waves is what gets illuminated by the inhabiter's beam of primordial consciousness. The inhabiter's primordial consciousness is not the artificial intelligence of created conscious states, rather it is pure conscious awareness. In this way the kind or critical words heard by the inhabiter while thinking, are echoes of its primordial consciousness and what the echoes say to the inhabiter is based on its subjective interpretation. An inhabiter that knows it is not the pawn neither hopes for success nor fears failure; and this sort of confidence comes about from the inhabiter's consistent choice in the AI of the psyche altruistic.[11]

Though it is true that every signal received by the Inhabiter's focal point of dark energy from the universe's dark energy substrate layer will be gathered and adulterated by the normal matter computer brain, the cries of animals and human speech emerge from signals with connected origins. Every pawn that appears within the psyche came about because of the connection the inhabiter has with other focal points of dark matter, all morphed from the same dark energy that facilitates entanglement. All the voices heard by an inhabiter are orchestrated by its primordial consciousness and all the choices made by the Inhabiter it did so while thinking with artificial intelligence. It is only the dark matter inhabiter and not the normal matter computer brain that that is given the choice to think with the artificial intelligence of the psyche self-centered or the psyche altruistic. Thereby it is the inhabiter that will subjectively interpret what it hears in the voices of its choices that are echoed back to it.

APPENDIX B: PSYCHE MATRIX, PSYCHE SELF-CENTERED, PSYCHE ALTRUISTIC

In order to understand the Wachowskis' Matrix movies in the way that complements this book, then an analogy must be made where the Matrix is understood as being a human psyche, which is the created consciousness of a human being that is experienced by an inhabiter. In the book, there are two psyches (psyche self-centered and psyche altruistic) that are created simultaneously. However, because the inhabiter can only use the cognitive states of one psyche to think, it is aware of only one psyche at any given moment. In the movie, there is just one psyche actually named, and it is called the Matrix; however, there are two psyches actually depicted in the movie, which is similar to the book. At the beginning of the *Matrix* movie, it is primarily the psyche self-centered that is displayed, but as the movie progresses, which includes the movie's two sequels, *Matrix Reloaded* and *Matrix Revolution*, the psyche altruistic is what is being portrayed. In the Matrix movies, the dark matter inhabiter is not seen, and while you are watching the movie, imagine that you are witnessing the psyche that the inhabiter has chosen to experience its momentary reality.

In the book, the living being that the inhabiter perceives in both psyches and from which it derives a sense of self while interacting with normal matter is called the pawn. Physically, the pawn looks the same in the psyche self-centered and the psyche altruistic. However, in the *Matrix* movie, the pawn physically looks very different. Neo is the pawn of the psyche altruistic, and Smith is the pawn of the psyche self-centered. Although Smith is not actually transformed into the pawn of the psyche self-centered until Neo jumps into Smith's body in the hallway scene at the end of the first Matrix movie. Thus the reader is reminded that even though the Matrix movies can be used as an aid in helping one understand the complexities of reality, the movies will not exactly correspond with the book or current scientific theory. Additionally, the psyche altruistic is not portrayed in the movies as it is in the book due to the continued killing and violence that would not balance the equation of interaction between an inhabiter and a normal matter computer brain, nor benefit the nonseparable universe. However, the sacrifices the pawn Neo is characterized as making in situations that appear within the movie fit with that of the pawn of the psyche altruistic described in the book.

There are definitions and short explanations of the specific terms that appear in the Matrix correlations and are located at the beginning of each chapter, starting with chapter 2 in the book. A short description of the movie scene and a line from the selected scene has been noted. The descriptions and lines are meant to clue the reader that is familiar with the movies or is just watching them for the first time as to the scene that is discussed in the correlation.

The Matrix

The Matrix scene: Neo and Trinity meet face-to-face for the first time in a nightclub.
Line reference from selected scene:

Trinity: Please just listen. I know why you're here, Neo ... and sit at your computer ...[1]

Life and the Aura of Perpetual Impermanence: The Dark Matter Inhabiter, the Pawn, and the Normal Matter Computer Brain correlation:
Every inhabiter has their own questions that emerge within the cognitive states of the psyche that an inhabiter chooses and thinks with. However, when an inhabiter selects the psyche altruistic, they are using their primordial consciousness derived from their focal point of dark energy in a way that will find an accurate answer. The real question each inhabiter seeks to answer when choosing the psyche altruistic is how in their unique way they might best cooperate with normal matter so as to balance the

equation of interaction. As each inhabiter participates in a round of the game of life while embodied within a normal matter computer brain, they will symbolically contribute a focal point of information that collectively will be used to help answer the universal question. This is the reason why dark energy morphed into individual dark matter inhabiters. There are myriad ways that dark matter inhabiters and normal matter computer brains occupy the universe while interacting, but does the synergy between these two energies result in universal bliss? If not, would it be best if these two energies coexisted in the universe in their noninteractive forms, as ambient normal energy that glides free on the universe's dark energy substrate layer surface, and might this eliminate suffering and result in universal bliss?

The Matrix scene: Neo meets Agent Smith for the first time after being captured, and this meeting precedes Agent Smith's transformation to the pawn of the psyche self-centered.
Line reference from selected scene:

Agent Smith: As you can see, we've had our eye on you for some time now, Mr. Anderson.[2]

Life and the Aura of Perpetual Impermanence: The Dark Matter Inhabiter, the Pawn, and the Normal Matter Computer Brain correlation:
Every inhabiter embodied within a normal matter computer brain begins its very first round in the game of life making an initial choice of the psyche self-centered. This means every inhabiter will begin its interaction with normal matter immersed in the cognitive states of the psyche self-centered and will believe itself to be the pawn. The dark matter inhabiter is not a living being, though it thinks that it is and will experience physical sensations through the pawn's body and will believe that it is perceiving the phenomenal world through the pawn's eyes.

The inhabiter of the movie believes that it is Neo that abides by societal constraints and pays its taxes while secretly hacking into the Matrix. This is not a completely off-base assumption made by the inhabiter because its choices do indeed enable its imprint strings that will send the signals that will alter or keep status quo the programming used by the normal matter computer brain's meta-algorithm. However, the fact that the inhabiter is not Neo but believes it is means cognitively this inhabiter is indeed living multiple lives, and when the normal matter computer brain computes for impermanence, this particular round of the game of life ends. Thus the pawn Neo does not have a future; however, given the fact that the dark matter inhabiter and the normal matter computer brain are energies that do not die, they do have a future that is determined by the inhabiter's participation in the game of life.

The Matrix scene: Neo meets Morpheus for the first time.
Line reference from selected scene:

Morpheus: At last. Welcome, Neo. As you no doubt have guessed, I am Morpheus.[3]

Life and the Aura of Perpetual Impermanence: The Dark Matter Inhabiter, the Pawn, and the Normal Matter Computer Brain correlation:

The rules to the game of life are that the dark matter inhabiter while embodied within a normal matter computer brain must use the cognitive states of the psyche to think and make choices. While the inhabiter's beam of primordial consciousness with a central region penetrates the layered electromagnetic radiation, it will be surrounded by the electromagnetic waves of both psyches that it will illuminate to create the phenomenal world that surrounds the pawn Neo. When the inhabiter chooses the psyche self-centered, then it will perceive the pawn symbolically taking the blue pill, and when the inhabiter chooses the psyche altruistic, then the pawn is seen by the inhabiter taking the red pill.

The world as portrayed in the three-dimensional hologram of the psyche self-centered is so simplified that it dummies down the complexities of the universe. The most dangerous misconception held by an inhabiter that habitually chooses the psyche self-centered is that it is in control of its life; this is an incorrect assumption. For as long as the dark matter inhabiter continues to choose the psyche self-centered, it will be a slave to its ignorance.

Each inhabiter begins to choose the psyche altruistic for unique reasons, albeit, when the inhabiter makes this selection, life and the universe emerge within this psyche in a way that loosens the inhabiter's grip on its tightly held beliefs. The inhabiter, having chosen this psyche, will have opened dimensions to cognitive states that initially will feel, from the inhabiter's point of view, like it is tumbling down a rabbit hole into a world that is more fascinating than the world portrayed in *Alice in Wonderland*.

The Matrix scene: Neo has taken the red pill.
Line reference from selected scene:

Morpheus: Time is always against us.[4]

Life and the Aura of Perpetual Impermanence: The Dark Matter Inhabiter, the Pawn, and the Normal Matter Computer Brain correlation:

The exact time when the normal matter computer brain computes for impermanence is unknown to the dark matter inhabiter. For however long the inhabiter has been immersed in the cognitive states of the psyche self-centered means the inhabiter has work to do, and it's in a race with brain death to disable its ego-clinging imprint strings before the inevitable computation for impermanence.

The Matrix scene: Neo going through the initial stages of being liberated from the Matrix.
Line reference from selected scene:

Morpheus: Have you ever had a dream, Neo, that you were so sure was real.[5]

Life and the Aura of Perpetual Impermanence: The Dark Matter Inhabiter, the Pawn, and the Normal Matter Computer Brain correlation:

While thinking with the cognitive states of the psyche self-centered, the inhabiter is experiencing a nonlucid dream or a nonlucid nightmare. When thinking with the cognitive states of the psyche altruistic, the inhabiter is experiencing a lucid dream or a lucid therapeutic nightmare. The inhabiter will not perceive the real world and wake up until it has managed to disable all of its ego-clinging imprint strings and normal matter and dark matter are liberated from their interaction.

The Matrix scene: Neo has left his pod in the Matrix and is on Nebuchadnezzar.
Line reference from selected scene:

Neo: Why do my eyes hurt?[6]

Life and the Aura of Perpetual Impermanence: The Dark Matter Inhabiter, the Pawn, and the Normal Matter Computer Brain correlation:

Because a dark matter inhabiter actually has no eyes, only enabled imprint strings and the primordial consciousness of its focal point of dark energy, an inhabiter does not directly "see" itself or the normal matter computer brain. What the inhabiter perceives is what it thinks with the cognitive states of a psyche, and this inhabiter in the movie still believes itself to be the pawn Neo.

The Matrix scene: Morpheus talks to Neo in his room on the Nebuchadnezzar.
Line reference from selected scene:

Morpheus: You believe it's the year 1999 when in fact it's closer to 2199.[7]

Life and the Aura of Perpetual Impermanence: The Dark Matter Inhabiter, the Pawn, and the Normal Matter Computer Brain correlation:

Time is an illusory concept that exists only within the psyche.

The Matrix scene: Morpheus and Neo in the white room (loading program) with chair.
Line reference from selected scene:

Morpheus: This is the construct. It's our loading program.[8]

Life and the Aura of Perpetual Impermanence: The Dark Matter Inhabiter, the Pawn, and the Normal Matter Computer Brain correlation:

Neo is the pawn of the psyche altruistic, and each inhabiter will perceive a unique pawn that is the avatar of its direct interaction they are having with the normal matter computer brain. Although pawns have no actual reality and emerge only within the psyche, inhabiters will derive a sense of

238

self from the pawn that begins approximately eighteen months from the time they begin inhabiting a normal matter computer brain.[9] When the inhabiter chooses one of the psyches to experience its momentary reality, computer brain programs are loaded and running, the cognitive states of that psyche define situations as real. As the inhabiter's choices enable and disable its imprint strings that send signals to the computer brain and result in alternative programming or allow the computer brain to process data according to its nature, the situations perceived by the inhabiter are real in their consequence.

The Matrix scene: Morpheus explains to Neo about the Matrix.
Line reference from selected scene:

Morpheus: We have only bits and pieces of information …[10]

Life and the Aura of Perpetual Impermanence: The Dark Matter Inhabiter, the Pawn, and the Normal Matter Computer Brain correlation:
The dark matter inhabiter does not know when exactly it began interacting with the normal matter computer brain and if this life represents its first round in the game of life. However, before its transformation to become a dark matter inhabiter, it was a part of the universe's dark energy substrate layer. While embodied within a normal matter computer brain, the inhabiter must think with what could be perceived as artificial intelligence, as the universe's dark energy does not think with cognitive states. When some inhabiters think with the cognitive states of the psyche self-centered and believe themselves to be the pawn, they will marvel at their own intelligence, which is a display of their ignorance. Unlike the movie, the dark matter inhabiter began its interaction with absolutely no malicious intention toward normal matter; in fact, it was a heroic gesture of compassion, and the inhabiter immersed itself in a vulnerable situation.

The Matrix scene: Morpheus explains to Neo about the Matrix (continued).
Line reference from selected scene:

Morpheus: When the Matrix was first built …[11]

Life and the Aura of Perpetual Impermanence: The Dark Matter Inhabiter, the Pawn, and the Normal Matter Computer Brain correlation:

Every inhabiter has a focal point of dark energy as part of its configuration that is attached to the universe's dark energy substrate layer. Until all inhabiters have completed their odyssey of discovery, the interaction with normal matter will continue. However, there are spread throughout the universe enlightened inhabiters that could have managed to liberate themselves but chose to remain in conditioned existence so as to be of benefit to inhabiters, normal matter computer brains, and the nonseparable universe. Though enlightened inhabiters are committed to remaining embodied by a normal matter computer brain for as long as necessary to be of benefit, each inhabiter must balance their own equation of interaction with normal matter.

The Matrix scene: Morpheus explains to Neo about the Matrix (continued). Line reference from selected scene:

Morpheus: Get some rest, you're going to need it. [12]

Life and the Aura of Perpetual Impermanence: The Dark Matter Inhabiter, the Pawn, and the Normal Matter Computer Brain correlation:

It takes time and commitment for an inhabiter to transform itself from one that consistently chooses the psyche self-centered to the alternative by choosing the psyche altruistic.

The Matrix scene: Neo fights Morpheus on the Nebuchadnezzar. Line reference from selected scene:

Morpheus: This is a sparring program … [13]

Life and the Aura of Perpetual Impermanence: The Dark Matter Inhabiter, the Pawn, and the Normal Matter Computer Brain correlation:

Every time the inhabiter selects the psyche altruistic, it is training itself through cognitive states that create its alternative and beneficial perception. The psyche altruistic is made of electromagnetic waves created by enabled non-ego-clinging imprint strings that bend and tap into the inhabiter's focal point of dark energy. Although enabled non-ego-clinging imprint strings receive guidance from the inhabiter's focal point of dark energy and their oscillations are changed, the inhabiter must abide by some of the computer brain's rules that govern neural functionality. The inhabiter must free itself from its unrewarding reactions and responses of choosing the psyche self-centered that cause it to cognitively perceive security that does not exist.

The Matrix scene: Back on the Nebuchadnezzar after Neo falls in the jump program.
Line reference from selected scene:

Neo: I thought it wasn't real. [14]

Life and the Aura of Perpetual Impermanence: The Dark Matter Inhabiter, the Pawn, and the Normal Matter Computer Brain correlation:

Though the inhabiter is not the pawn, when choosing a psyche, the inhabiter will experience the illusion that through the character's body, the inhabiter smells, tastes, touches, hears, and sees phenomenon. Yet in reality, these are nothing more than three-dimensional holograms that emerge at the subtle conscious level of the layered electromagnetic radiation. All phenomena are electromagnetic waves containing information and are illuminated by the inhabiter using its primordial consciousness. When the computer brain computes for impermanence, it will be temporarily freed and dispersed throughout the universe as normal energy, until drawn into engagement by the gravitational force produced by an inhabiter's enabled imprint strings.

The Matrix scene: Busy street program.
Line reference from selected scene:

Morpheus: The Matrix is a system ... [15]

Life and the Aura of Perpetual Impermanence: The Dark Matter Inhabiter, the Pawn, and the Normal Matter Computer Brain correlation:

When other living beings emerge within the psyche self-centered, the inhabiter will misapprehend the situation and believe that these character represent others that are fundamentally different from the inhabiter, which is not the case. Every living being other than the pawn that appears within the psyches is created either when an inhabiter's focal point receives quantum signals from the universe's dark energy substrate layer or the computer brain receives signals from the network of normal matter computer brains it is connected to. The universe's dark energy orchestrates the phenomenon of entanglement and thereby facilitates communication between inhabiters. Once a signal is received by an inhabiter's focal point of dark energy, its enabled imprint strings will incorporate this information into the electromagnetic waves of the psyche self-centered and the psyche altruistic. When an inhabiter selects a psyche, this will disable and enable the inhabiter's imprint strings, and the imprint strings that are left enabled will send signals to the computer brain. Thus the initial quantum signal sent by the universe's dark energy substrate layer to an inhabiter is now a part of the normal matter computer brain's programming. The computer brain's meta-algorithm will transform this signal's information into neural synchrony and will further modulate and integrate it before sending it to circuitry located throughout the computer brain and network of normal matter the computer brain is connected to. There is no certainty when this information might later be sent by the normal matter computer brain through signals back to the inhabiter's enabled imprint strings that transform them into electromagnetic waves of the psyches. Thus when the inhabiter perceives other living beings, it cannot know if its focal point is actively receiving signals from the universe's dark energy substrate layer or if this living being represents the computer brain's recreation and modifying of a previous event. What the inhabiter can be sure of is that it is not the only dark matter inhabiter in the universe and that many inhabiters are lost in a nonlucid dream of the psyche self-centered. These lost inhabiters will feel threatened by the alternative reality viewpoint that appears within the psyche altruistic.

The Matrix scene: Neo approaches Cypher on the Nebuchadnezzar. Line reference from selected scene:

Cypher: Whoa, Neo. You scared the bejeezus out of me.[16]

Life and the Aura of Perpetual Impermanence: The Dark Matter Inhabiter, the Pawn, and the Normal Matter Computer Brain correlation:

What allows the dark matter inhabiter to understand the electrical pattern language of the normal matter computer brain is the transformation of this neural synchrony to electromagnetic waves by its enabled imprint strings. When the electromagnetic waves are illuminated with primordial consciousness derived from the inhabiter's focal point of dark energy, the inhabiter is able to decipher the computer brain's language while thinking with the cognitive states of a psyche.

The Matrix scene: In the mess hall on the Nebuchadnezzar.
Line reference from selected scene:

Tank: Here you go, buddy. Breakfast of champions.[17]

Life and the Aura of Perpetual Impermanence: The Dark Matter Inhabiter, the Pawn, and the Normal Matter Computer Brain correlation:

Only the inhabiter, not the computer brain, thinks with cognitive states. The normal matter computer brain is unaware of the inhabiter's existence and does not perceive smells, taste food, hear sounds, see forms, or feel textures; rather, it responds to degrees of neural synchrony.

The Matrix scene: Neo in the car with Trinity and Morpheus headed to see the Oracle.
Line reference from selected scene:

Neo: I used to eat there.[18]

Life and the Aura of Perpetual Impermanence: The Dark Matter Inhabiter, the Pawn, and the Normal Matter Computer Brain correlation:

The inhabiter that has habitually chosen the psyche self-centered has been cognitively adrift in a nonlucid dream while misapprehending what it is and its existence. What constitutes the phenomena that the inhabiter thinks are its memories actually did not happen, as they were inaccurately perceived by the inhabiter. When the inhabiter chooses the psyche altruistic, it will be experiencing a lucid dream and not actual reality; symbolically, however, the inhabiter can formulate an accurate relative viewpoint of reality.

The Matrix scene: Neo meets one of the orphans at the Oracle's apartment. Line reference from selected scene:

Spoon boy: Do not try and bend the spoon … [19]

Life and the Aura of Perpetual Impermanence: The Dark Matter Inhabiter, the Pawn, and the Normal Matter Computer Brain correlation:

The inhabiter's fundamental error that causes it to habitually choose the psyche self-centered is its belief that it is the pawn. When the inhabiter consistently makes the alternative choice of the psyche altruistic and thereby enables its non-ego-clinging imprint strings, the inhabiter slowly realizes the truth: the independently existing human being that the inhabiter thinks of as "self" is an illusion within the psyches.

The Matrix scene: Neo meets the Oracle for the first time. Line reference from selected scene:

Oracle: So, what do you think?[20]

Life and the Aura of Perpetual Impermanence: The Dark Matter Inhabiter, the Pawn, and the Normal Matter Computer Brain correlation:

When the inhabiter believes itself to be the pawn of the psyche self-centered, the inhabiter wants all the living beings around it to confirm its importance. When the inhabiter chooses the psyche altruistic, over time the inhabiter grasps the significance of the pawn as the avatar that allows its direct interaction with normal matter. Wanting to balance the equation of interaction and understanding the implications of a nonseparable universe means the inhabiter is only interested in perceiving the pawn engaged in altruistic words and actions. The gift each inhabiter has is choice because there are two psyches created simultaneously by its enabled imprint strings.

The Matrix scene: Lafayette Hotel. Line reference from selected scene:

Neo: Whoa, déjà vu. [21]

Life and the Aura of Perpetual Impermanence: The Dark Matter Inhabiter, the Pawn, and the Normal Matter Computer Brain correlation:

In reality, déjà vu moments come about because both the psyche self-centered and psyche altruistic are created by enabled imprint strings that transform the computer brain's data. There are scenes within the psyche altruistic that are created specifically to train the inhabiter by recreating previous scenarios designed to afford the inhabiter the opportunity to make a beneficial choice and disable rather than enable its ego-clinging imprint strings.

The Matrix scene: Agents have captured Morpheus and they are trying to break into his mind in the office.
Line reference from selected scene:

Agent Smith: I'd like to share a revelation ... [22]

Life and the Aura of Perpetual Impermanence: The Dark Matter Inhabiter, the Pawn, and the Normal Matter Computer Brain correlation:

Inhabiters with an abundance of ego-clinging imprint strings enabled, having made the psyche self-centered their modus operandi, will believe rewarding themselves to be their highest priority.

The Matrix scene: Neo on the phone.
Line reference from selected scene:

Neo: I know you're out there. [23]

Life and the Aura of Perpetual Impermanence: The Dark Matter Inhabiter, the Pawn, and the Normal Matter Computer Brain correlation:

An inhabiter's world may be completely turned upside down as they change the manner by which they interact with the normal matter computer brain and choose the psyche altruistic. But the inhabiter can take comfort, although not in illusory certainty and security—instead, in the knowledge that it is one choice closer to liberation. The real universe has no rules and controls, nor limits to love, compassion, or bliss, and in this universe, anything is possible. However, to reach this destination, the inhabiter has to replicate this universe in the psyche altruistic while inhabiting a normal matter computer brain and living within the limits of conditioned existence.

The Matrix Reloaded

The Matrix Reloaded scene: The machines are digging.
Line reference from selected scene:

Niobe: These geotherms confirm …[24]

Life and the Aura of Perpetual Impermanence: The Dark Matter Inhabiter, the Pawn, and the Normal Matter Computer Brain correlation:

Only in the psyches are there countries, a planet where things are owned and where weapons of mass destruction exist. However, in the nonseparable universe, there are inhabiters communicating via entanglement, normal matter computer brains connected and communicating via a network of normal matter, and dark matter inhabiters embodied within normal matter computer brains. When interacting with each other, inhabiters and normal matter computer brains are unpredictable, and the many versions of the psyches perceived by inhabiters are impermanent. Inhabiters misapprehending their conditioned existence have not understood the very real threat that exists because the universe is nonseparable and has a dark energy substrate layer that facilitates entanglement. As a result of the nonseparable universe where there is communication between inhabiters that experience conditioned existence lost in nonlucid dreams and nightmares of the psyche self-centered, all of this could potentially culminate to produce a period of destruction of monumental proportions in the form of a group nightmare of world war and nuclear annihilation.

The Matrix Reloaded scene: Temple gathering.
Line reference from selected scene:

Morpheus: Zion! Hear me! It is true, what many of you have heard.[25]

Life and the Aura of Perpetual Impermanence: The Dark Matter Inhabiter, the Pawn, and the Normal Matter Computer Brain correlation:

Every inhabiter is aware that brain death is an event that they cannot avoid. Some inhabiters fear it, while other inhabiters that have consistently chosen the psyche altruistic prepare for it so they are ready for their transformation. Each round of the game of life includes a period of destruction that is marked by a final conflagration uniquely experienced by each inhabiter.[26] What is important to consider is that the psyches contain electromagnetic waves of both sense impressions and mental events. This means that the inhabiter will experience mental events as three-dimensional moving holograms but will also feel physical sensations. Which is to say that the period of destruction is an event that includes images and physical sensation that will be experienced differently by each inhabiter, dependent on its choices and population of enabled imprint strings. This makes the period of destruction a critical event in the game of life.

The Matrix Reloaded scene: Care for some company?
Line reference from selected scene:

Councilor Hamann: Almost no one comes down here …[27]

Life and the Aura of Perpetual Impermanence: The Dark Matter Inhabiter, the Pawn, and the Normal Matter Computer Brain correlation:

The normal matter computer brain does not think and has no idea that the inhabiter exists. Nonetheless, each inhabiter is embodied within a normal matter computer brain that computes for impermanence, and when exactly this event will take place is unknown to the inhabiter. Many inhabiters spread throughout the universe simply live their life of conditioned existence in the nonlucid dream of psyche self-centered, never taking time to use cognitive states to learn much more than what they need to satiate their wants and needs. Thus, discoveries in many areas of science are conveniently ignored, and the spiritual is not cognitively absorbed or linked but rather thought by these inhabiters as being unrelated to science. However, logically, this sort of thinking makes little sense since science and spirituality both emerge within the cognitive states of the psyche that grants each inhabiter the ability to think. There are inhabiters that become lost in the black hole of despair while experiencing the psyche self-centered as a nonlucid nightmare and assume suicide is an escape where life ends and they along with their pain disappear. This is a dangerous mistaken assumption by the inhabiter because in a universe consisting of interacting energy, there are not actual living beings that are born, and there are no living beings that die except within the psyche. As the total sum of the universe's mass and energy does not change,[28] nothing can start to exist or cease to exist, and where there is no violation of the principle of the conservation of mass-energy, there are only transformations.[29] Although there are many mysteries in the universe, while embodied and as a consequence of perceiving the psyche altruistic, the inhabiter can formulate a relative view of reality that is accurate, and if lost in a black hole, this psyche provides the means by which an inhabiter can liberate itself from darkened cognitive states. All it takes is for the inhabiter to consistently make an alternative choice of the psyche altruistic.

The Matrix Reloaded scene: Seraph takes Neo to the Oracle.
Line reference from selected scene:

Neo: These are back doors, aren't they?[30]

Life and the Aura of Perpetual Impermanence: The Dark Matter Inhabiter, the Pawn, and the Normal Matter Computer Brain correlation:
Think of the code described by Seraph to be electrical notes and the tumblers as neural synchrony. The eighty-six billion[31] neurons of the brain of a normal matter computer brain configured to produce data enabled imprint strings will use to make a human psyche are firing randomly and independently. The paired structure of normal matter known as the claustra detect, modulate, and integrate the pulse propagating from not one but billions of neurons. Neurons firing together can synchronize to produce the same electrical note. The more synchronized the firing of neurons in a particular cortical area is, the more synchronized the note will be and the more likely it will be to gain the claustra's attention.[32]

The Matrix Reloaded scene: Neo and the Oracle in the courtyard.
Line reference from selected scene:

Oracle: Well, come on. I ain't gonna bite ya.[33]

Life and the Aura of Perpetual Impermanence: The Dark Matter Inhabiter, the Pawn and the Normal Matter Computer Brain correlation:

Although there are no actual living beings in the universe, only within the psyches, sometimes when electromagnetic waves with incorporated transformed neurotransmitter data are illuminated by the central region of the inhabiter's beam of primordial consciousness, living beings emerge within the psyches as menacing characters, such as the movie's Smith, Agents, Sentinels, Merovingian, Trainman, and so on, and phenomenal disasters such as machines digging into Zion. However dangerous or upsetting the inhabiter perceives these manifestations to be, in actuality, they are not caused by an independent entity or force external to its focal point of dark energy and the layered electromagnetic radiation.

A metaphorical role model is produced within the psyche altruistic via quantum signals sent by an enlightened inhabiter and creates a living being, which in the movie is the Oracle. The Oracle is meant to guide the dark matter inhabiter into the metaphysical realm of reality, and this inhabiter is drawn to the metaphorical role model because of the connection it feels to the complex set of attitudes and self-regulatory behaviors that the inhabiter can identify with and use as behavioral guides. Yet, without knowing it, this inhabiter is experiencing the identity principle, because when it follows the Oracle, it is actually connecting with its primordial consciousness in a very meaningful way. The psyche altruistic, therefore, includes ordinary people, but the character that was created with an enlightened inhabiter's signal is in no way ordinary.[34] This character represents pure consciousness of an enlightened inhabiter that made a decision to remain embodied within a normal matter computer brain to be of benefit. The means by which an inhabiter is able to distinguish the metaphorical role model as being in the psyche altruistic is similar to the Oracle's activities that are for the sake of other living beings.[35] Nothing this character does is to become famous, rich, or important, and its accomplishments are purely motivated by the desire to benefit others.[36]

The Matrix Reloaded scene: Smiths and Neo fight in the courtyard.
Line reference from selected scene:

Smith: Then you're aware of it.[37]

Life and the Aura of Perpetual Impermanence: The Dark Matter Inhabiter, the Pawn, and the Normal Matter Computer Brain correlation:

In the courtyard movie scene, there is a stream of Smiths depicted, and this fits with the actual way the normal matter computer brain creates data for the pawn. The computer brain's meta-algorithm will detect neural synchrony, modify timing of synchronized pulses, and serially sequence agents. Materializing from this activity will be three different versions of the computer brain's I character that serve a different purpose as the computer brain uses them to generate different stages within the information processing cycle. To compute for the here and now, the meta-algorithm uses the core-I[38] character, and when historical data is accessed, the autobiographic-I[39] character emerges within the psyches as the inhabiter's imprint strings transform the computer brain's data and fabricate this character. The appearance of one I character with a body that moves and speaks in the three-dimensional hologram of a psyche at the subtle conscious level of the layered electromagnetic radiation is an illusion.

Similar to the movie, there are many versions of the pawn created within both psyches that correspond to the moment the computer brain's meta-algorithm imperceptibly vacillated between computing for the core-I and autobiographical-I character. As the meta-algorithm must maintain internal states for the normal matter computer brain to continue to function as a unit, it will correlate synchrony of biologic information first.[40] The proto-I character is the primitive and internal milieu presence programming within all versions of the I character.[41] Cognizing transformed neurotransmitters as feelings from this starting point, the inhabiter's awareness of the psyche self-centered proceeds with a commanding urge for survival and to maintain stability. When Smith is depicted as "hacking" a living being by the spread of black liquid in the movie, this symbolically depicts an inhabiter's immersion and what happens to it when choosing the psyche self-centered.

Dark energy created ego-clinging imprint strings for the purpose of allowing the inhabiter an immersive experience into the world of normal matter as it is being sequenced by the normal matter computer brain. When choosing the psyche self-centered, the inhabiter will have an "insider" experience and will perceive the world as normal matter creates it. Ego-clinging imprint strings will create conscious states of the psyche self-centered, according to the computer brain's agenda, which is to benefit the brain's neuronal circuitry. The conscious states of the psyche self-centered are predatory in nature, as the concept of self is a priority of the inhabiter, having chosen this psyche. When an inhabiter chooses the psyche self-centered, for an undisclosed but variable amount of time, it becomes immersed and loses the ability to perceive without duality. Instead, the inhabiter redefines its life purpose by making the psyche self-centered its modus operandi, and to varying degrees, the inhabiter perceives the pawn of the psyche self-centered exhibiting predatory and narcissistic behavior.

The Matrix Reloaded scene: Restaurant of the Merovingian.
Line reference from selected scene:

Morpheus: You know why we are here.[42]

Life and the Aura of Perpetual Impermanence: The Dark Matter Inhabiter, the Pawn, and the Normal Matter Computer Brain correlation:
   The normal matter computer brain can be thought of as a trafficker of information because it receives and shares data with the network of computer brains spread throughout the universe that it is connected to. Exactly how data is relayed throughout the network of connected computer brains is perplexing, as there is no assuredness to the direction of information flow or how it will be received.
   Indeed, choice is a luxury afforded to the inhabiter, not the normal matter computer brain, and only when there are sufficient numbers of its enabled non-ego-clinging imprint strings does the inhabiter have the power to control its perception. The degree to which its non-ego-clinging imprint strings are enabled corresponds to the inhabiter's ability to freely choose its actions without being influenced by the computer brain directing itself. Free, however, does not mean unconstrained, since the inhabiter's actions are free but determined by its enabled imprint strings.[43]
   The inhabiter has the opportunity to weigh its options before responding and choosing a psyche by which to experience momentary reality. However, the inhabiter can forfeit this option by quickly responding

before perceiving the alternative response within the cognitive states of the other psyche. Albeit, it would be wise if the inhabiter was careful before making its choices due to the proto presence of created conscious states. The underlying primeval feeling will be experienced by the inhabiter as it perceives the pawn of the psyche self-centered seeking food, fluids, fun, and reproductive opportunities.[44]

The Matrix Reloaded scene: Keymaker explains the task. The Keymaker's explanation correlates with the book and the period of destruction when normal matter begins its computation for impermanence.
Line reference from selected scene:

Keymaker: There's a building. Inside this building there's a level where no elevator can go …[45]

Life and the Aura of Perpetual Impermanence: The Dark Matter Inhabiter, the Pawn, and the Normal Matter Computer Brain correlation:
   Think of the normal matter computer brain as the building referred to by the Keymaker. How normal matter folds around the inhabiter creates the shape and structure of the normal matter computer brain. The computer brain is an integrated electrical system, and normal matter will transmit information as waves of electrical patterns. The interclaustral pathway is the computer brain's explicit layer of circuitry where the meta-algorithm alters and selects the best neural synchrony based on rules; its actions are therefore highly regulated. The neural nets within the cerebral cortex make up the computer brain's implicit layer of circuitry and are arranged as a hybrid architecture of interconnected networks.[46]
   When an agent contains synchrony that does not fit well within the computer brain's version of the universe, it is perceived as a threat, and the meta-algorithm modifies the timing of its processing. Rapidly and automatically, the meta-algorithm responds to the threatening agent with a directive that activates circuitry within a neuronal net and may be the catalyst for the computation for impermanence.[47] If the brain stem interprets the proto-synchrony as conveying a threat with a low probability of reward, then it utilizes its connections to widespread circuitry to send out its neurotransmitter programs.[48] Although the computer brain has a

large repertoire of neurotransmitter programs, those created by the brain stem are particularly important because this structure sets the associative rules that the meta-algorithm follows when processing information. Norepinephrine is the neurotransmitter program that causes the "bomb" described by the Keymaker in the movie, and when transformed by enabled imprint strings at high levels, the inhabiter will experience feelings that precede sense impressions or mental events that are associated with anxiety and stress.

While inhabiting a normal matter computer brain that is configured to produce data for a human psyche, the claustra's meta-algorithm and the inhabiter's focal point of dark energy are not directly connected. Instead, the meta-algorithm and the inhabiter's focal point of dark energy with a surface of imprint strings occupy two separate and distinct regions within the normal matter computer brain. Enabled ego-clinging imprint strings create the psyche self-centered, and enabled non-ego-clinging imprint strings create the psyche altruistic via four steps and during two consecutive rotations of these steps. During the second rotation, the oscillations of enabled imprint strings increase, and this causes the electromagnetic radiation of two psyches to coil upon itself to produce layers. The funnel-shaped electromagnetic has a narrowed base that widens with increasing layers of electromagnetic radiation. The base of the layered electromagnetic radiation contains the least amount of electromagnetic waves and represents the gross conscious level. As the layered electromagnetic radiation gets higher, it widens and contains more electromagnetic waves representing the subtle conscious level. The gross conscious level contains electromagnetic waves of sense impressions, and the subtle conscious level contains electromagnetic waves of mental events. Because the narrowed base of the funnel-shaped electromagnetic radiation does not directly make contact with the surface of the focal point of dark energy but levitates just above it, a space is created between the inhabiter's dark energy focal point's surface and layered electromagnetic radiation. This space represents the extremely subtle conscious level that is devoid of electromagnetic waves of sense impressions and mental events.

Thus the layered electromagnetic radiation contains the levels described by the Keymaker in the "building" of the normal matter computer brain. As the inhabiter's data transformers, enabled imprint strings are the symbolic builders of doors that, when opened, create the inhabiter cognitive states. The inhabiter's focal point of dark energy is the figurative key maker and with primordial consciousness possesses all the keys to the doors created by enabled imprint strings. During the period of destruction, as the many layers of the brain's interconnected networks are crashing, the inhabiter's enabled imprint strings will continue to create electromagnetic waves via two rotations of the four-step sequence. The inhabiter's failsafe during this event is to choose the psyche altruistic and, with its beam of primordial consciousness with a central region, enter the space of the extremely subtle conscious level. The inhabiter may or may not have just over five minutes to do this; the inhabiter cannot stop the computer brain's computation for impermanence, but it does not need to. During the last moment before brain death,[49] if the inhabiter is perceiving the extremely subtle conscious level, then it will experience bliss, luminosity, and nonconceptuality.[50]

The Matrix Reloaded scene: Morpheus explains …
Line reference from selected scene:

Morpheus: All of our lives we have fought this war. Tonight I believe we can end it.[51]

Life and the Aura of Perpetual Impermanence: The Dark Matter Inhabiter, the Pawn and the Normal Matter Computer Brain correlation:
The purpose of life that is everlasting is about transformation to be of benefit to the nonseparable universe.

The Matrix Reloaded scene: Niobe struggles with doubt.
Line reference from selected scene:

Morpheus: What is it, Niobe?[52]

Life and the Aura of Perpetual Impermanence: The Dark Matter Inhabiter, the Pawn, and the Normal Matter Computer Brain correlation:
No single theory will predict all the incalculable and unique versions of illusory reality observed by inhabiters within the psyche self-centered and the psyche altruistic. The number of inhabiters exceeds that of existing stars,[303] and each makes choices that are not predetermined but rather

impacted by their enabled and disabled imprint strings. This does not mean, however, that inhabiters need shy away from assembling evidence and discerning their motivations before choosing the psyche by which to experience momentary reality. It is the choice of the psyche altruistic and the enabling of non-ego-clinging imprint strings that will manipulate the environment of a normal matter computer brain and guide inhabiters to a deeper understanding of the nonseparable universe. Although the psyche altruistic may be profoundly different from what an inhabiter is used to, it is only the inhabiter that can free its embodied mind from fear, doubt, and disbelief contained within the cognitive states of the psyche self-centered.

The Matrix Reloaded scene: Neo meets the Architect.
Line reference from selected scene:

Neo: Who are you?[53]

Life and the Aura of Perpetual Impermanence: The Dark Matter Inhabiter, the Pawn, and the Normal Matter Computer Brain correlation:
There is a fundamental rule as it relates to the equation of interaction between dark matter inhabiters and normal matter computer brains: the energy that possesses primordial consciousness must use this gift skillfully to create a harmonious interaction before dark matter and normal matter are enduringly freed from their interaction. Only the inhabiter with its focal point of dark energy as part of its configuration has primordial consciousness yet also has enabled and disabled imprint strings. Imprint strings are a part of the inhabiter and are not annihilated by the actions of the opposite string; to do so would be self-harming to the inhabiter. However, imprint strings keep a tally of the interaction between the inhabiter and the normal matter computer brain when they are enabled and disabled. Complexity and uncertainty are aspects built into the equation of interaction and life embodied within a normal matter computer brain, and this fact is not initially grasped by an inhabiter as it begins its odyssey of discovery.

Many inhabiters have participated in an abundant number of games of life in countless diversely configured normal matter computer brains. Embodied within a normal matter computer brain, the inhabiter experiences a psyche that deceptively displays what the inhabiter understands to be its life. It would be more accurate if the inhabiter adopted the viewpoint that each psyche is analogous to a normal matter computer brain/imprint-string-based simulated environment. Within the psyches, interacting energies are made to appear solid, and where there is unpredictability, predictable order appears. Unlike in the movie, however, if the inhabiter is entering into another round in the game of life, it does not cognitively select anything; instead, the gravitational force of its enabled imprint strings draw normal energy to a region of the universe where it is transformed to normal matter that will embody the inhabiter as a computer brain.

Each inhabiter is a morphed focal point of the same universe's dark energy substrate layer that has explored its interactions with normal matter in a myriad of complex scenarios facilitated by entanglement. When an inhabiter completes its odyssey of discovery and no longer interacts with normal matter, the universe's dark energy substrate layer is aware of this and keeps track of interactions between dark matter inhabiters and normal matter computer brains but only as physical attributes,[54] not illusory appearance of individuals or limitations based on color, country, species, or time.

The Matrix Revolutions

The Matrix Revolutions scene: Oracle's home. The Oracle has changed her appearance and does not look as she did in *The Matrix* movie.
Line reference from selected scene:

Trinity: Who are you?[55]

Life and the Aura of Perpetual Impermanence: The Dark Matter Inhabiter, the Pawn, and the Normal Matter Computer Brain correlation:
Fully aware of its presence and that of normal energy in the universe, dark energy understood the ramifications before beginning its interaction with normal energy. In order to overcome any barriers that might hinder its efforts to live harmoniously with normal energy, dark energy ascertained that it must engage normal energy to attain vital information. Nonetheless, there is much more dark energy in the universe than normal energy;

thus, if dark energy engaged normal energy during one interaction, allegorically, normal energy would be overpowered and swallowed up by the experience. Therefore, the universe's dark energy substrate layer designed the interaction with normal energy in a way that accommodated the inequivalence of power. However, dark energy understood that while interacting with normal matter as dark matter, it is only the inhabiter that will suffer as it struggles to think with its cognitive states and experiences loss, aging, and death.

As the Oracle in the movie was made by the quantum signals from an enlightened inhabiter, her comments reveal an enlightened inhabiter's motivations, which are deeply rooted in unfaltering commitment to remain in a conditioned state of existence for as long as it takes to benefit the nonseparable universe.

The Matrix Revolutions scene: Train station when Neo meets Ram-Kendra and he speaks of love.
Line reference from selected scene:

Ram-Kendra: I know only what I need to know.[56]

Life and the Aura of Perpetual Impermanence: The Dark Matter Inhabiter, the Pawn, and the Normal Matter Computer Brain correlation:
The inhabiter perceiving itself as the pawn and immersed in the cognitive states of the psyche self-centered has a profound attachment to only those to whom it feels connected, yet the word itself is nothing but the inhabiter's interpretation of neural synchrony generated by a computer brain that does not feel the same connection that the word implies.

The Matrix Revolutions scene: Oracle's home. Neo confronts the Oracle and wants answers to his questions.
Line reference from selected scene:

Oracle: Remember what you were like when you first walked through my door?[57]

Life and the Aura of Perpetual Impermanence: The Dark Matter Inhabiter, the Pawn, and the Normal Matter Computer Brain correlation:

The metaphorical role model, the Oracle guides the inhabiter in cognitive experiences that include discipline, concentration, and knowledge, and these will be presented within scenes of the psyche altruistic in ways that will be easy for the inhabiter to assimilate[58] if studied and reflected upon. When the inhabiter questions why things appear as they do and diligently searches for answers, a thought emerges that there is something wrong with its perception of the world. If the inhabiter remains open to what it learns, then it begins to grasp the relative and infinite dimensions of its reality and with a widened perspective. The inhabiter has the power to change its inaccurate beliefs and redefine what is real. Each inhabiter has a focal point of dark energy as part of its configuration; thus, it is power with grace that extends beyond conditioned existence and reaches to the nonseparable universe.

The Matrix Revolutions scene: Oracle's home. Neo confronts the Oracle and wants answers to his questions (continued).
Line reference from selected scene:

Oracle: Everything that has a beginning, has an end.[59]

Life and the Aura of Perpetual Impermanence: The Dark Matter Inhabiter, the Pawn, and the Normal Matter Computer Brain correlation:

Smith is the pawn of the psyche self-centered, and Neo is the pawn of the psyche altruistic. The illusory phenomena within the psyches have no true existence, and every round of the game of life has a beginning and an end.

The Matrix Revolutions scene: Oracle's home. Smith confronts the Oracle.
Line reference from selected scene:

Smith: The great and powerful Oracle.[60]

Life and the Aura of Perpetual Impermanence: The Dark Matter Inhabiter, the Pawn, and the Normal Matter Computer Brain correlation:

The greatest power of the metaphorical role model and Oracle is that this character possesses compassion, empathy, enlightened intelligence, and creativity. When the inhabiter chooses the psyche self-centered, it believes the way to overcome threats ultimately is to destroy them, as depicted in the movie by the pawn Smith. However, when the inhabiter chooses the psyche altruistic, it will learn from the metaphorical role model, the Oracle, that thinks of solutions outside of the dominance/destruction box. Every inhabiter has an invisible Oracle within its configuration and is its focal point of dark energy. To connect with limitless ability for compassion and creativity, the inhabiter need only choose the psyche altruistic created by its non-ego-clinging imprint strings tapping into its focal point of dark energy.

The Matrix Revolutions scene: Logos. Trinity is mortally wounded.
Line reference from selected scene:

Neo: What? Oh no ...[61]

Life and the Aura of Perpetual Impermanence: The Dark Matter Inhabiter, the Pawn, and the Normal Matter Computer Brain correlation:

The living beings that appeared as three-dimensional holograms at the subtle conscious level of the layered electromagnetic radiation created an illusion within the psyches of relationships between the pawn and others. Hence, relationships do not mean interactions between distinct, intrinsically existing objects, such as living beings, but rather symbolically represent a network of normal matter computer brains and focal points of dark matter that communicate with other inhabiters while attached to the universe's dark energy substrate layer.

The Matrix Revolutions scene: Machine City. Neo makes a deal with Deus Ex Machina.
Line reference from selected scene:

Neo: I only have come to say what I want to say.[62]

Life and the Aura of Perpetual Impermanence: The Dark Matter Inhabiter, the Pawn, and the Normal Matter Computer Brain correlation:

The inhabiter in the movie at this point has disabled an abundance of its ego-clinging imprint strings and is thinking more with the cognitive states of the psyche altruistic.

The Matrix Revolutions scene: Somewhere on a street.
Line reference from selected scene:

Smith: Can you feel it, Mr. Anderson?[63]

Life and the Aura of Perpetual Impermanence: The Dark Matter Inhabiter, the Pawn, and the Normal Matter Computer Brain correlation:

Inhabiters' fundamental error in perception—an error that motivates their actions and was the catalyst for their unrewarding responses in choosing the psyche self-centered—was their misidentification of themselves as pawns. Ergo, when these inhabiters enabled their ego-clinging imprint strings, they were unable to profoundly bend their perception concerning what they thought they knew about themselves and the phenomenal world before the period of destruction and brain death.

When an inhabiter chooses the psyche altruistic, difficult cognitive themes of the psyche self-centered will be recreated, giving the inhabiter the opportunity to disable its ego-clinging imprint strings it enabled through its previous nonbeneficial choices. Inhabiters that used their primordial consciousness to guide themselves from the abyss of their inaccurate perceptions and were able to think with the cognitive states of the psyche altruistic will have made progress in reemergence through discovery learning before the period of destruction.[64] Thereby, these inhabiters do not perceive brain death as terminal but rather the making of their transformation. Every round of the game of life has a beginning and an end of an interaction between a normal matter computer brain and a dark matter inhabiter whose participation has been tallied by its enabled imprint strings.

The Matrix Revolutions scene: Oracle and the Architect talk somewhere in a park, and Oracle reunites with Sati and Seraph.
Line reference from selected scene:

Oracle: Well now, ain't this a surprise.[65]

Life and the Aura of Perpetual Impermanence: The Dark Matter Inhabiter, the Pawn, and the Normal Matter Computer Brain correlation:

Neo is the pawn and will not emerge again within the psyche altruistic. As the movie depicted light shooting out from Neo's eyes and mouth as well as Smith's, a round in the game of life was concluded. Accurately, however, the Oracle describes the inhabiter's journey continuing, and for this inhabiter it will be as an enlightened inhabiter. Having radically eradicated its difficulties in interacting with normal matter and balanced the equation of interaction through its participation, this inhabiter grasped the deeper meaning to its existence and the complexity of the nonseparable universe. As an enlightened inhabiter, it will indeed communicate with other inhabiters spread throughout the universe.

GLOSSARY

acetylcholine: A neurotransmitter that when released by some neurons of the computer brain and transformed by enabled imprint strings, the inhabiter will experience either a pleasant, unpleasant, or neutral feeling that precedes a sense impression or mental event that involves memory, sleep, the pawn's muscle activation, and other nervous-system functions.[1]

agent/algorithm: A self-contained set of step-by-step operations the meta-algorithm uses to perform tasks such as data processing.[2]

amygdala: One of the four subcortical nuclei in each cerebral hemisphere that is part of the limbic system in some normal matter computer brains.[3]

anterior tract: A pathway located within the normal matter computer brain's implicit layer of circuitry.

atom of cognition: Transformed agents that function as the neural correlate to the inhabiter's created conscious states.

autobiographical-I character:[4] One of three optimally intelligent agents, this character has a progressing history.

basal ganglia (or basal nuclei): A group of subcortical nuclei that are strongly interconnected with the cerebral cortex, thalamus, brain stem, and several other areas in normal matter computer brains of vertebrates, including humans.[5]

binding: A phenomenon that takes place when the normal matter computer brain's claustra connect separate pieces of sensory information while responding to degrees of synchrony from firing neurons.[6]

black hole: A celestial object that emerges within the psyches and has a gravitational field so strong that light cannot escape it, and potentially was created in the collapse of a very massive star.[7] Yet black holes are allegorical and are created by the inhabiter using its enabled imprint strings so as to conceptualize its very existence while inhabiting a normal matter computer brain.

blueprint: A record of the claustra's input and actions stored in clarification spaces located within the computer brain's implicit layer of circuitry.[8]

brain stem: A primordial normal matter computer brain structure that connects the spinal cord with the forebrain and cerebrum.[9]

cerebral cortex (pallium): Convoluted surface of the cerebrum of the normal matter computer brain that functions chiefly to collect and coordinate sensory and motor information.[10]

cerebral hemisphere: Two convoluted halves of the computer brain's cerebrum.[11]

circuit: A neuronal pathway in the normal matter computer brain along which electrical and chemical signals travel.[12]

claustra: A paired structure located beneath the cerebral cortex on the left and right side of the brain of the normal matter computer brain.[13]

cognition: Cognitive mental processes of the inhabiter.[14]

computer brain: Normal matter that embodies an inhabiter that is composed of neurons and supporting structures that integrates sensory information in controlling autonomic function, in coordinating and directing correlated motor responses.[15]

core-I character:[16] One of three optimally intelligent agents and does not include algorithms of historical perspective. When this character is perceived by the inhabiter within a psyche, it is the pawn.

corpus callosum: A band of 200 million nerve fibers uniting the cerebral hemispheres of the normal matter computer brain.[17]

dark energy: Most prevalent energy present in the universe and has natural intelligence that is devoid of created conscious states.

dark matter: A form dark energy uses when interacting with normal matter.

discovery principle: Constituting principle used to promote the inhabiter's ability to learn while exploring the world it perceives within the psyches.[18]

dopamine: A neurotransmitter released by some neurons of the normal matter computer brain when computing for reward. When transformed by enabled imprint strings, the inhabiter will experience it as a pleasant, unpleasant, or neutral feeling that precedes a sense impression or mental event.[19]

ego-clinging imprint string: Negative magnetic monopole that transforms the computer brain's data to electromagnetic waves as directed by normal matter.

electromagnetic radiation: Energy in the form of electromagnetic waves that contains the psyche self-centered and the psyche altruistic.[20]

emotions: Differing frequencies of neural synchrony experienced by the inhabiter and not by the normal matter computer brain. There are six primary emotions: happiness, sadness, fear, anger, surprise, and disgust[21] and four secondary emotions: embarrassment, jealousy, guilt, and pride.[22]

electromagnetic waves: Waves of energy produced by synchronized oscillations from electric and magnetic fields.[23] Moving charges create magnetic forces, and moving magnets create electric forces.

elementary particle (fundamental particle): An elementary (or fundamental) particle whose substructure is unknown; thus, it is unknown whether it is composed of other particles.[24]

essential experience: A phenomenal experience that manifests differently depending on the inhabiter's choice of a psyche to experience momentary reality and the population of its enabled imprint strings.[25]

explicit layer of circuitry: Circuitry located in the interclaustral pathway of the normal matter computer brain.

forebrain: Anterior of three primary divisions of the developing vertebrate normal matter computer brain that includes the cerebral hemispheres, the thalamus, and the hypothalamus.[26]

genetic marker:[27] Assigned by the brain stem to "proto-" information.

glutamate: A prominent neurotransmitter, and when released by the normal matter computer brain, it enhances the synchronicity of the pulse propagating between two neurons.[28] When accumulated in a synaptic cleft and it spills over to adjacent synapses in other circuits, cross-talking between neurons occurs.[29] When transformed by enabled imprint strings, the inhabiter will experience a pleasant, unpleasant, or neutral feeling that precedes a sense impression and mental event but with an increase in volume of the computer brain's data, created by neuronal impulses throughout the brain's circuitry.[30]

gray matter: Neural tissue, especially of the normal matter computer brain's spinal cord and brain, that contains nerve cell bodies as well as nerve fibers.[31]

hybridization:[32] Term used to describe the computer brain's hybrid architecture made of interconnected networks that allow normal matter to accommodate neural connectivity within a limited space.

hypothalamus: A structure that lies beneath the thalamus on each side of normal matter's brain[33] and has been recognized for its importance in creating the computer brain's data for the inhabiter's emotional behavior.[34]

I character: The computer brain's optimally intelligent agent that acts as a centralized control of the claustra's processing for perception, cognition, and action.[35] When this character is perceived by the inhabiter within a psyche, it is the pawn.

identity principle: A guiding principle meant to encourage the inhabiter to make beneficial choices.[36]

immersion: A term used to describe an inhabiter immersed in the cognitive states of the psyche self-centered.[37]

implicit layer of circuitry: Made of neural nets[38] within the cerebral cortex of the computer brain and arranged as a hybrid architecture[39] of interconnected networks of normal matter.

imprint strings: Magnetic monopoles of two types, ego-clinging and non-ego-clinging, that transform the computer brain's neural synchrony into electromagnetic waves, which allows the dark matter inhabiter to interact with normal matter. Imprint strings can be enabled or disabled but not destroyed.

inhabiter: A focal point of dark matter, which is a focal point of the universe's dark energy with a surface of imprint strings.

intelligence: The ability of an inhabiter, while thinking with cognitive states, to utilize its primordial consciousness in a way that allows the inhabiter to acquire skills and knowledge while interacting with the normal matter computer brain.[40] The ability a normal matter computer brain has to be plastic with the capacity for its circuitry to be molded or altered, especially when receiving signals from enabled non-ego-clinging imprint strings.[41]

interclaustral pathway: A pathway located within a region of the normal matter computer brain called the corpus callosum.[42]

lateral tract: A normal matter tract that relays electrical patterns from the auditory cortex to the computer brain's claustra.[43]

law of cause and effect: A relation between a cause and its effect between regularly correlated events while the dark matter inhabiter interacts with the normal matter computer brain.[44]

local-type electromagnetic field: A phenomenon where a local type of electromagnetic field is produced when an inhabiter's focal point of dark energy is receiving signals from the universe's dark energy substrate layer.

macroscopic universe: An illusory demarcation created within the psyches that is perceived by an inhabiter allowing it to conceptualize the small dimensions of the universe.

medial tract: A normal matter pathway within the computer brain's implicit layer of circuitry that relays electrical patterns that contains data for the inhabiter's conscious perception of emotion and goal-driven behavior.[45]

meta-algorithm: A phenomenon that emerges in the interclaustral pathway when claustra share their information with one another and correlate the separate synchronized neural activity[46] within the computer brain's different sensory circuits.[47]

motor neuron: A nerve cell (neuron) whose cell body is located in the spinal cord and whose fiber (axon) projects outside the spinal cord to directly or indirectly control muscles made of normal matter.[48]

neocortex: The large six-layered dorsal region of the cerebral cortex of a normal matter computer brain that is configured to produce the psyche of a mammal.[49]

networks: Neural networks made of normal matter's interconnected circuitry of neurons.[50]

neural nets: Linked computer brain circuits that when layered form networks of neural nets that compute and store information to operate as templates of associative memory.[51]

neuron: One of the many billions of electrical cells that collectively form the fabric of the normal matter computer brain.[52]

neurotransmitters: Chemical programs the normal matter computer brain uses to modulate the activity of its neurons, allowing them to communicate and transmit information. When transformed by enabled imprint strings, the inhabiter while thinking experiences them as pleasant, unpleasant, or neutral feelings.

non-ego-clinging imprint string: Positive magnetic monopole that transforms the computer brain's data to electromagnetic waves according to the guidance it receives from the inhabiter's focal point of dark energy.

norepinephrine (NE) or noradrenalin: A neurotransmitter that when released by some neurons of the computer brain and transformed by enabled imprint stings the inhabiter will experience a pleasant, unpleasant, or neutral feeling that precedes a sense impression or mental event with arousal, focused attention, increased restlessness, anxiety, heart rate, and blood pressure; it also triggers the release of glucose from energy stores and increases blood flow to the pawn's skeletal muscles.[53]

normal energy: One of the two types of energies in the universe that resides undecidedly free of impermanent states of existence until it is transformed to its interactive form normal matter.

normal matter: The visible form normal energy uses when interacting with dark matter.[54] **nucleus:** A cluster of cell bodies of neurons in the central nervous system, located deep within the normal matter computer brain's cerebral hemispheres and brain stem.[55]

occipital lobe: The normal matter computer brain's posterior area that specializes in creating data used for the inhabiter's vision.

photon: An elementary particle, the quantum of the electromagnetic field including electromagnetic radiation.[56]

posterior tract: A normal matter pathway located within the computer brain's implicit layer of circuitry that relays information to the claustra from the areas of normal matter that create data for vision.[57] When the data is transformed by enabled imprint strings, the inhabiter while thinking will have the conscious perception of form.[58]

proto-I[59] character: One of three optimally intelligent agents that signifies the normal matter computer brain's existence in an integrated form and is the foundation on which all other versions of I character are created.[60]

quantum: A quantum (plural: quanta) is the minimum amount of any physical entity involved in an interaction.[61]

reality: The quality or state of being real as defined by an inhabiter while thinking with the cognitive states of a psyche.[62]

serotonin: A neurotransmitter of the normal computer brain that the inhabiter will experience as a pleasant feeling that precedes a mental event or physical sensation; an immersed inhabiter cognizes that the mental event or sense impression is the cause of its well-being and happiness.[63]

small-world topology: A term used to describe the normal matter computer brain's implicit layer of circuitry where its cortical areas are connected directly or through one or two intermediate areas.[64]

spinal cord: A long, thin, tubular bundle of normal matter in the form of nervous tissue and support cells that extends from the brain stem to the lumbar region of the vertebral column. The brain and spinal cord together make up the computer brain's central nervous system (CNS).[65]

270

substrate layer of the macroscopic universe: Made of dark energy in its two different forms (noninteractive form—dark energy, interactive form—dark matter).

superior tract: One of the five different normal matter pathways within the computer brain's implicit layer of circuitry that relays electrical patterns from sensory and motor areas[66] to create the data that when transformed by enabled imprint strings allows the inhabiter to experience the conscious perceptions of smell, touch, and taste.[67]

thalamus: A paired normal matter structure that is joined at the midline and located near the center of the brain of a computer brain. It relays data in many ways, but the exact way in which thalamic circuitry controls information flow to the cortex is unknown.[68]

vertebrate: A normal matter computer brain that is configured to produce the psyche of animals (e.g., mammals, birds, reptiles, amphibians, and fishes).[69]

virtual world: A world created with the normal matter computer brain's data that has been transformed into electromagnetic waves of information by enabled imprint strings. When the inhabiter illuminates the electromagnetic waves with its beam of primordial consciousness, the virtual world emerges within the psyche self-centered and the psyche altruistic.

ENDNOTES FOR LIFE AND THE AURA OF PERPETUAL IMPERMANENCE: THE DARK MATTER INHABITER, THE PAWN, AND THE NORMAL MATTER COMPUTER BRAIN

1. "Decision-Making Process," accessed July 27, 2016, http://umass.edu./.
2. Ibid.
3. "Dark Energy, Dark Matter," NASA Science, accessed July 24, 2016, http://science.nasa.gov/astrophysics/focus-areas/what-is-dark-energy.
4. Ibid.
5. Dilgo Khyentse, *The Collected Works of Dilgo Khyentse Volume One* (Boston: Shambhala Publications, Inc., 2010), 129.
6. *A Brief History of Time*, directed by Errol Morris (Tokyo Broadcasting Systems, Channel Four Films (aka Film Four International, aka Channel 4 TV), Anglia Television Ltd., National Broadcasting Company (NBC), 1992), film.
7. Khyentse, *The Collected Works of Dilgo Khyentse Volume Two*, 28.
8. "Dark Energy, Dark Matter," NASA Science.
9. Ibid.
10. Khyentse, *The Collected Works of Dilgo Khyentse Volume Two*, 122.
11. Morris, *A Brief History of Time*.
12. Khyentse, *The Collected Works of Dilgo Khyentse Volume One*, 149.
13. Morris, *A Brief History of Time*.
14. "Dark Energy, Dark Matter," NASA Science.
15. Matthieu Ricard and Trinh Xuan Thuan, *The Quantum and the Lotus* (New York, New York: Three Rivers Press, 2001), 165.
16. Khyentse, *The Collected Works of Dilgo Khyentse Volume Two*, 31.
17. Khyentse, *The Collected Works of Dilgo Khyentse Volume One*, 128.
18. Ricard and Thuan, *The Quantum and the Lotus*, 26.

19. "Consciousness," Bernard J. Baars, Scholarpedia 10, no. 8 (2015): 2207. Accessed November 11, 2015. doi:10.4249/scholarpedia.2207.

20. "Dark Energy, Dark Matter," NASA Science.

21. "Freewill and Determinism," University of Central Lancashire, Center for Professional Ethics, accessed June 20, 2016, http://uclan.ac.uk.

22. Khyentse, *The Collected Works of Dilgo Khyentse Volume One*, 134.

23. "Dark Energy, Dark Matter," NASA Science.

24. "Decision-Making Process."

25. "Measuring and Defining the Experience of Immersion in Games," Charlene Jennett et al., UCL Interaction Centre University College London Remax House, retrieved July 25, 2016, http://www-users.cs.york.ac.uk.

26. Ibid.

27. Dilgo Khyentse, *The Collected Works of Dilgo Khyentse Volume Three* (Boston: Shambhala Publications, Inc., 2010), 95.

28. "Decision-Making Process."

29. "Making Sense of String Theory," Brian Greene, YouTube, February 2005, http://ted.com.

30. Morris, *A Brief History of Time*.

31. Greene, "Making Sense of String Theory."

32. Khyentse, *The Collected Works of Dilgo Khyentse Volume One*, 134.

33. "Dark Energy, Dark Matter," NASA Science, accessed July 24, 2016.

34. Morris, *A Brief History of Time*.

35. Greene, "Making Sense of String Theory."

36. Khyentse, *The Collected Works of Dilgo Khyentse Volume One*, 128.

37. Morris, *A Brief History of Time*.

38. "The Quest to Understand Consciousness," Antonio Damasio, TED video, filmed March 2011, posted December 19, 2011, http://www.ted.com/.

39. Ibid.

40. Morris, *A Brief History of Time*.

41. "How Many Neurons Make a Human Brain? Billions Fewer than We Thought," James Randerson, *The Guardian*, February 28, 2012, accessed March 16, 2016, http://www.theguardian.com.

42. "A Brain in a Supercomputer," Henry Markram, TED video, filmed July 2009, posted April 15, 2009, http://www.ted.com/.

43. Ibid.

44. Andrea D. Grabovac, Mark A. Lau and Brandilyn Willett, "Mechanisms of Mindfulness: A Buddhist Psychological Model," Springer Science+Business Media, LLC. (2011).

45. "Dark Energy, Dark Matter," NASA Science.

46. Khyentse, *The Collected Works of Dilgo Khyentse Volume One*, 313.

47. Greene, "Making Sense of String Theory."

48. Jennett et al., "Measuring and Defining the Experience of Immersion in Games."

49. Morris, *A Brief History of Time.*

50. Jennett et al., "Measuring and Defining the Experience of Immersion in Games."

51. Ibid.

52. Khyentse, *The Collected Works of Dilgo Khyentse Volume Two*, 31.

53. Markram, "A Brain in a Supercomputer."

54. Khyentse, *The Collected Works of Dilgo Khyentse Volume Two*, 313.

55. *The Matrix*, directed by Andy Wachowski and Larry Wachowski (Burbank, CA: Warner Brothers, 1999), DVD.

56. Morris, *A Brief History of Time.*

57. John W. Folkins et al., "Enhancing the Therapy Experience Using Principles of Video Game Design," *American Journal of Speech-Language Pathology*, February 25, 2016, 111–21.

58. Khyentse, *The Collected Works of Dilgo Khyentse Volume Two*, 228.

59. Scott Rogers, *Level Up!: The Guide to Great Video Game Design* (West Sussex, United Kingdom: Wiley, 2010).

60. Khyentse, *The Collected Works of Dilgo Khyentse Volume One*, 134.

61. Ibid.

62. Ibid.

63. "Hawking Radiation," (n.d.), Wikipedia, retrieved September 22, 2016 from http://wikipedia.org/wiki/Hawking_radiation.

64. "Freewill and Determinism," University of Central Lancashire, Center for Professional Ethics.

65. "Emergence: The Unconscious Toscanini of the Brain," Ellen Gordon, Jad Abumrad (producers), February 18, 2005 [audio podcast] retrieved from www.radiolab.org.

66. Randerson, "How Many Neurons Make a Human Brain? Billions Fewer than We Thought."

67. "Emergence: The Unconscious Toscanini of the Brain," Ellen Gordon, Jad Abumrad (producers).

68. Ibid.

69. Ibid.

70. John Smythies, Lawrence Edelstein, and Vilayanur Ramachandran, "Hypotheses Relating to the Function of the Claustrum," *Frontiers in Integrative Neuroscience* 6 (August 2012): 53.

71. Ibid.

72. D. Milardi et al., "Cortical and Subcortical Connections of the Human Claustrum Revealed in Vivo by Constrained Spherical Deconvolution

Tractography," *Cerebral Cortex* 25, no. 2 (February 2015): 406–14. doi:10.1093/cercor/ bht231.

73. Ibid.

74. Khyentse, *The Collected Works of Dilgo Khyentse Volume Two*, 112.

75. Milardi et al., "Cortical and Subcortical Connections of the Human Claustrum Revealed in Vivo by Constrained Spherical Deconvolution Tractography," 406–14.

76. Khyentse, *The Collected Works of Dilgo Khyentse Volume Two*, 112.

77. Milardi et al., "Cortical and Subcortical Connections of the Human Claustrum Revealed in Vivo by Constrained Spherical Deconvolution Tractography," 406–14.

78. Khyentse, *The Collected Works of Dilgo Khyentse Volume Two*, 112.

79. Milardi et al., "Cortical and Subcortical Connections of the Human Claustrum Revealed in Vivo by Constrained Spherical Deconvolution Tractography," 406–14.

80. Khyentse, *The Collected Works of Dilgo Khyentse Volume Two*, 112.

81. Milardi et al., "Cortical and Subcortical Connections of the Human Claustrum Revealed in Vivo by Constrained Spherical Deconvolution Tractography," 406–14.

82. Khyentse, *The Collected Works of Dilgo Khyentse Volume Two*, 112.

83. Milardi et al., "Cortical and Subcortical Connections of the Human Claustrum Revealed in Vivo by Constrained Spherical Deconvolution Tractography," 406–14.

84. Smythies, Edelstein, and Ramachandran, "Hypotheses Relating to the Function of the Claustrum," 53.

85. Ibid.

86. Baars, "Consciousness."

87. Ibid.

88. Goertzel, "Artificial Intelligence."

89. Baars, "Consciousness."

90. Ibid

91. "Neural Correlates of Consciousness," Mormann Florian and Christof Koch, Scholarpedia 2, no.12(2007):1740. Accessed August 20, 2016. doi:1.4249/scholarpedia.1740.

92. Ibid.

93. Markram, "A Brain in a Supercomputer."

94. Ben Goertzel, "Artificial Intelligence," *Scholarpedia* 10 no. 11 (2015): 31847, accessed August 7, 2016. doi: 10.4249/scholarpedia.31847.

95. Goertzel, "Artificial Intelligence."

96. Goertzel, "Artificial Intelligence."

97. Smythies, Edelstein, and Ramachandran, "Hypotheses Relating to the Function of the Claustrum," 53.

98. Ibid.

99. Ibid.

100. Ibid.

101. Ibid.

102. "Neural Correlates of Consciousness," Mormann Florian and Christof Koch, Scholarpedia 2, no.12(2007):1740. Accessed August 20, 2016. doi:1.4249/scholarpedia.1740.

103. Baars, "Consciousness."

104. Smythies, Edelstein, and Ramachandran, "Hypotheses Relating to the Function of the Claustrum," 53.

105. Baars, "Consciousness."

106. Ibid.

107. Ibid.

108. Goertzel, "Artificial Intelligence."

109. F. C. Crick and C. Koch, "What Is the Function of the Claustrum?" *Philos. Trans. R. Soc. Lond. B Biol. Sci.* 360, (2005). 1271–1279.

110. Smythies, Edelstein, and Ramachandran, "Hypotheses Relating to the Function of the Claustrum," 53.

111. Ibid.

112. Ibid.

113. A. A. Ghazanfar and C. E. Schroeder, "Is the Neocortex Essentially Multisensory?" *Trends Cogn. Sci.* 10, (2006): 278–285.

114. Ibid.

115. Markram, "A Brain in a Supercomputer."

116. Smythies, Edelstein, and Ramachandran, "Hypotheses Relating to the Function of the Claustrum," 53.

117. Ibid.

118. Ibid.

119. Ibid.

120. Goertzel, "Artificial Intelligence."

121. Goertzel, "Artificial Intelligence."

122. Smythies, Edelstein, and Ramachandran, "Hypotheses Relating to the Function of the Claustrum," 53.

123. Ibid.

124. Ibid.

125. J. E. Bogen, "The Callosal Syndromes," in *Clinical Neuropsychology*, K. M. Heilman and E. Valenstein, eds. (New York: Oxford University Press, 1993), 337–407.

126. Smythies, Edelstein, and Ramachandran, "Hypotheses Relating to the Function of the Claustrum," 53.

127. U. Faghihi and S. Franklin, "The LIDA Model as a Foundational Architecture for AGI," in *Theoretical Foundations of Artificial General Intelligence*, P. Wang and B. Goertzel, eds. (Paris: Atlantis Press, 2012), 105–123.

128. Ibid.

129. "Neural Correlates of Consciousness," Mormann Florian and Christof Koch, Scholarpedia 2, no.12(2007):1740. Accessed August 20, 2016. doi:1.4249/scholarpedia.1740.

130. Smythies, Edelstein, and Ramachandran, "Hypotheses Relating to the Function of the Claustrum," 53.

131. Masuda and Aihara, "Spatiotemporal Spike Encoding of a Continuous External Signal," 1599–1628.

132. Smythies, Edelstein, and Ramachandran, "Hypotheses Relating to the Function of the Claustrum," 53.

133. Smythies, Edelstein, and Ramachandran, "Hypotheses Relating to the Function of the Claustrum," 53.

134. Howard Gardner, *Frames of Mind: The Theory of Multiple Intelligences* (New York: Basic Books, 1983).

135. Goertzel, "Artificial Intelligence."

136. Crick and Koch, "What Is the Function of the Claustrum?", 1271–1279.

137. Smythies, Edelstein, and Ramachandran, "Hypotheses Relating to the Function of the Claustrum," 53.

138. Ibid.

139. Florian and Koch, "Neural Correlates of Consciousness."

140. Ibid.

141. Markram, "A Brain in a Supercomputer."

142. Goertzel, "Artificial Intelligence."

143. Smythies, Edelstein, and Ramachandran, "Hypotheses Relating to the Function of the Claustrum," 53.

144. Ibid.

145. Goertzel, "Artificial Intelligence."

146. Ibid.

147. Smythies, Edelstein, and Ramachandran, "Hypotheses Relating to the Function of the Claustrum," 53.

148. John Holland, *Adaptation in Natural and Artificial Systems*, (Ann Arbor, Michigan: U. Michigan Press, 1975.) J. J. Hopfield, "Neural Networks and Physical Systems with Emergent Collective Computational Properties," *Proc. Nat. Acad. Sci.* (USA) 79, (1982) 2554–2558.

149. Goertzel, "Artificial Intelligence."

150. Ibid.

151. Barbara Hammer and Pascal Hitzler, eds., "Perspectives of Neural-Symbolic Integration," *Studies in Computational Intelligence*, Vol. 77. Springer, 2007.

152. Ibid.

153. Smythies, Edelstein, and Ramachandran, "Hypotheses Relating to the Function of the Claustrum," 53.

154. "Cognition and Emotion," Luis Pessoa, Scholarpedia 4, no 1: 4567. Accessed June 22, 2016. doi:10.4249/scholarpedia.4567.

155. Smythies, Edelstein, and Ramachandran, "Hypotheses Relating to the Function of the Claustrum," 53.

156. Ibid.

157. Baars, "Consciousness."

158. "A Theory of Universal Artificial Intelligence Based on Algorithmic Complexity," Marcus Hutter, 2000, arXiv:cs.AI/0004001.

159. Florian and Koch, "Neural Correlates of Consciousness."

160. "Neuroscientist Antonio Damasio Explains Consciousness," The Huffington Post, November 16, 2010, retrieved July 24, 2016, from http://www.huffingtonpost.com.

161. Ibid.

162. "Neuroscientist Antonio Damasio Explains Consciousness," The Huffington Post, November 16, retrieved July 24, 2016, from http://www.huffingtonpost.com.

163. Ibid.

164. Antonio Damasio, *The Feeling of What Happens. Body and Emotion in the Making of Consciousness* (New York: Harcourt, Inc., 1999), 136.

165. Ibid., 217.

166. Damasio, *The Feeling of What Happens. Body and Emotion in the Making of Consciousness*, 217.

167. "Neuroscientist Antonio Damasio Explains Consciousness," The Huffington Post.

168. Goertzel, "Artificial Intelligence."

169. Smythies, Edelstein, and Ramachandran, "Hypotheses Relating to the Function of the Claustrum," 53.

170. Goertzel, "Artificial Intelligence."

171. John Laird et al., "Claims and Challenges in Evaluating Human-Level Intelligent Systems," Proceedings of AGI-09, The Soar Cognitive Architecture, MIT Press, 2009.

172. "Neuroscientist Antonio Damasio Explains Consciousness," The Huffington Post.

173. Laird et al., "Claims and Challenges in Evaluating Human-Level Intelligent Systems."

174. J. J. Hopfield, "Neural Networks and Physical Systems with Emergent Collective Computational Properties," *Proc. Nat. Acad. Sci.* (USA) 79, (1982): 2554–2558.
175. Goertzel, "Artificial Intelligence."
176. Laird et al., "Claims and Challenges in Evaluating Human-Level Intelligent Systems."
177. Laird et al., "Claims and Challenges in Evaluating Human-Level Intelligent Systems."
178. Goertzel, "Artificial Intelligence."
179. Ibid.
180. Laird et al., "Claims and Challenges in Evaluating Human-Level Intelligent Systems."
181. Goertzel, "Artificial Intelligence."
182. Ibid.
183. Goertzel, "Artificial Intelligence."
184. Smythies, Edelstein, and Ramachandran, "Hypotheses Relating to the Function of the Claustrum," 53.
185. Damasio, "The Quest to Understand Consciousness."
186. *Fentanyl: The Drug Deadlier than Heroin*, VICE, published on July 22, 2016, accessed September 30, 2016.
187. Antonio Damasio, *The Feeling of What Happens. Body and Emotion in the Making of Consciousness* (New York: Harcourt, Inc., 1999), 49.
188. Pessoa, "Cognition and Emotion."
189. Damasio, *The Feeling of What Happens. Body and Emotion in the Making of Consciousness*, 125.
190. Goertzel, "Artificial Intelligence."
191. Damasio, *The Feeling of What Happens. Body and Emotion in the Making of Consciousness*, 323.
192. "Neuroscientist Antonio Damasio Explains Consciousness," The Huffington Post.
193. Damasio, *The Feeling of What Happens. Body and Emotion in the Making of Consciousness*, 217.
194. Damasio, *The Feeling of What Happens. Body and Emotion in the Making of Consciousness*, 217.
195. "Neuroscientist Antonio Damasio Explains Consciousness," The Huffington Post.
196. Hutter, "A Theory of Universal Artificial Intelligence Based on Algorithmic Complexity."
197. Goertzel, "Artificial Intelligence."
198. Smythies, Edelstein, and Ramachandran, "Hypotheses Relating to the Function of the Claustrum," 53.
199. Ibid.
200. Ibid.

201. Ibid.

202. Ibid.

203. "Emotional Memory," Joseph E. LeDoux, Scholarpedia 2, no. 7: (2007): 2698. Accessed June 22, 2016. doi:10.4249/scholarpedia.1806.

204. Pessoa, "Cognition and Emotion."

205. "Neural Basis of Emotions," Antonio Damasio, Scholarpedia 6, no. 3: 1804. Accessed June 22, 2016. doi: 10.4249/scholarpedia.1804.

206. Damasio, *The Feeling of What Happens. Body and Emotion in the Making of Consciousness*, 50.

207. Ibid., 51.

208. L. W. Swanson, *Brain Architecture: Understanding the Basic Plan* (New York: Oxford University Press, 2003).

209. L. W. Swanson, "Cerebral Hemisphere Regulation of Motivated Behavior," *Brain Res* 886 (2000):113–164.

210. Laird et al., "Claims and Challenges in Evaluating Human-Level Intelligent Systems."

211. Hopfield, "Neural Networks and Physical Systems with Emergent Collective Computational Properties," 2554–2558.

212. Swanson, "Cerebral Hemisphere Regulation of Motivated Behavior," 113–164.

213. Damasio, *The Feeling of What Happens. Body and Emotion in the Making of Consciousness*, 156.

214. Hutter, "A Theory of Universal Artificial Intelligence Based on Algorithmic Complexity."

215. Swanson, *Brain Architecture: Understanding the Basic Plan.*

216. Damasio, "The Quest to Understand Consciousness."

217. "Brainstem" (n.d.), in Wikipedia, retrieved September 4, 2016 from http:// en.wikipedia.org/wiki/Brainstem.

218. Damasio, "The Quest to Understand Consciousness."

219. "Neuroscientist Antonio Damasio Explains Consciousness," The Huffington Post.

220. Holland, "Adaptation in Natural and Artificial Systems." Hopfield, "Neural Networks and Physical Systems with Emergent Collective Computational Properties," 2554–2558.

221. F. C. Crick and C. Koch, "A Framework for Consciousness," *Nat. Neurosci.* 6 (2003), 119–126.

222. "Neuroscientist Antonio Damasio Explains Consciousness," The Huffington Post.

223. Goertzel, "Artificial Intelligence."

224. Ron Sun, *Duality of the Mind: A Bottom-Up Approach Toward Cognition* (Mahwah, NJ: Lawrence Erlbaum Associates, 2002).

225. Goertzel, "Artificial Intelligence."

226. LeDoux, "Emotional Memory."

227. Wachowski and Wachowski, *The Matrix*.

228. LeDoux, "Emotional Memory."

229. "Glutamate Receptor" (n.d.), in Wikipedia, retrieved September 4, 2016 from http://en.wikipeidia.org/wiki/Glutamate_receptor.

230. "Visualization of Glutamate as a Volume Transmitter," *J. Physiol.*, 2011 Feb 1;589 (Pt. 3):481–8. doi: 10.1113/jphysiol.2010.199539. Epub 2010 Nov 29.

231. Faghihi and Franklin, "The LIDA Model as a Foundational Architecture for AGI," 105–123.

232. Goertzel, "Artificial Intelligence."

233. Damasio, "Neural Basis of Emotions."

234. Goertzel, "Artificial Intelligence."

235. Damasio, *The Feeling of What Happens. Body and Emotion in the Making of Consciousness*, 136.

236. Holland, "Adaptation in Natural and Artificial Systems." Hopfield, "Neural Networks and Physical Systems with Emergent Collective Computational Properties," 2554–2558.

237. A. N. Kolmogorov, "Three Approaches to the Quantitative Definition of Information," *Problems of Information and Transmission*, 1(1) (1965):1–7

238. "Thermodynamic Equilibrium" (n.d.), in Wikipedia, retrieved January 19, 2017 from https://en.wikipedia.org/wiki/Thermodynamic_equilibrium.

239. "Electromagnetic Radiation," (n.d.), in Wikipedia, retrieved February 9, 2017 from https://en.wikipedia.org/wiki/Electromagnetic_radiation.

240. Ibid.

241. "Can brain waves interfere with radio waves," (n.d.), in Ask an Engineer, retrieved May 30, 2018 from https://engineering.mit.edu/engage/ask-an-engineer/can-brain-waves-interfere-with-radio-waves/

242. "Wireless power transfer," (n.d.), in Wikipedia, retrieved May 30, 2018 from https://en.wikipedia.org/wiki/Wireless_power_transfer.

243. Ibid.

244. "Black-Body Radiation" (n.d.), in Wikipedia, retrieved October 19, 2016 from https://en.wikipedia.org/wiki/Randomness.

245. Khyentse, *The Collected Works of Dilgo Khyentse Volume Three*, 569.

246. Ibid., 571.

247. "Quantum," (n.d.), in Wikipedia, retrieved February 5, 2017 from https://en.wikipedia.org/wiki/Quantum.

248. "Electromagnetic Radiation," Wikipedia.

249. B. Alan Wallace, *The Attention Revolution* (Somerville, MA: Wisdom Publications, Inc., 2006), 138.

250. Ibid., 137.

251. Ibid., 121.

252. Dennis Overbye, "No Escape from Black Holes? Stephen Hawking Points to a Possible Exit," *New York Times*. Last modified June 2016.https://www.nytimes.com/2016/06/07/science/stephen-hawking-black-holes.html?_r=0.

253. "Electromagnetic Radiation," Wikipedia.

254. Ibid.

255. Grabovac, Lau and Willett, "Mechanisms of Mindfulness: A Buddhist Psychological Model."

256. "Neuroscientist Antonio Damasio Explains Consciousness," The Huffington Post.

257. Grabovac, Lau and Willett, "Mechanisms of Mindfulness: A Buddhist Psychological Model."

258. Ibid.

259. Ibid.

260. "Electromagnetic Radiation," Wikipedia.

261. Ibid.

262. Ibid.

263. Ibid.

264. Wallace, *The Attention Revolution*, 137.

265. Greene, "Making Sense of String Theory."

266. Morris, *A Brief History of Time*.

267. Jennett et al., "Measuring and Defining the Experience of Immersion in Games."

268. "Acetylcholine" (n.d.), in Wikipedia, retrieved March 19, 2017 from https://en.wikipedia.org/wiki/Acetylcholine.

269. "Dopamine" (n.d.), in Wikipedia, retrieved March 19, 2017 from https://en.wikipedia.org/wiki/Dopamine.

270. "Norepinephrine" (n.d.), in Wikipedia, retrieved March 19, 2017 from https://en.wikipedia.org/wiki/Norepinephrine.

271. "Serotonin" (n.d.), in Wikipedia, retrieved March 19, 2017 from https://en.wikipedia.org/wiki/Serotonin.

272. "Glutamate Receptor" (n.d.), in Wikipedia, retrieved September 4, 2016 from http://en.wikipeidia.org/wiki/Glutamate_receptor.

273. "Visualization of Glutamate as a Volume Transmitter," *J Physiol.*, 2011 Feb 1;589 (Pt. 3):481–8. doi: 10.1113/jphysiol.2010.199539. Epub 2010 Nov 29.

274. Wachowski and Wachowski, *The Matrix*.

275. Morris, *A Brief History of Time*.

276. "Thomas theorem" (n.d.), in Wikipedia, retrieved May 31, 2018 from https://en.wikipedia.org/wiki/Thomas_theorem

277. Karl M. Kapp, "The Gamification of Learning and Instruction: Game-Based Methods and Strategies for Training and Education." (San Francisco: Wiley, 2012), 10.

278. Folkins et al., "Enhancing the Therapy Experience Using Principles of Video Game Design," 111–21.

279. Khyentse, *The Collected Works of Dilgo Khyentse Volume One*, 142.

280. Khyentse, *The Collected Works of Dilgo Khyentse Volume Two*, 107.

281. Damasio, *The Feeling of What Happens. Body and Emotion in the Making of Consciousness*, 50.

282. "Neuroscientist Antonio Damasio Explains Consciousness," The Huffington Post.

283. Jennett et al., "Measuring and Defining the Experience of Immersion in Games."

284. Damasio, *The Feeling of What Happens. Body and Emotion in the Making of Consciousness*, 127.

285. "Neuroscientist Antonio Damasio Explains Consciousness," The Huffington Post.

286. Ibid.

287. "Optic Nerve" (n.d.), in Wikipedia, retrieved October 19, 2016 from http://en.wikipedia.org/wiki/Optic_nerve.

288. Damasio, "The Quest to Understand Consciousness."

289. Ibid.

290. Wachowski and Wachowski, *The Matrix*.

291. Khyentse, *The Collected Works of Dilgo Khyentse Volume One*, 323

292. Wachowski and Wachowski, *The Matrix*.

293. Khyentse, *The Collected Works of Dilgo Khyentse Volume Three*, 32.

294. Khyentse, *The Collected Works of Dilgo Khyentse Volume One*, 142.

295. Khyentse, *The Collected Works of Dilgo Khyentse Volume Two*, 107.

296. Ibid., 111.

297. Ibid.

298. Damasio, *The Feeling of What Happens. Body and Emotion in the Making of Consciousness*, 55.

299. Khyentse, *The Collected Works of Dilgo Khyentse Volume Three*, 32.

300. M. Csikszentmihalyi, *Flow: The Psychology of Optimal Experience* (New York: Harper and Row, 1990).

301. Ibid.

302. "Black-Body Radiation" (n.d.), in Wikipedia, retrieved October 19, 2016 from https://en.wikipedia.org/wiki/Randomness.

303. Khyentse, *The Collected Works of Dilgo Khyentse Volume Two*, 317.

304. Ibid.

305. Jennett et al., "Measuring and Defining the Experience of Immersion in Games."

306. Folkins et al., "Enhancing the Therapy Experience Using Principles of Video Game Design," 111–21.

307. John E. Calamari (PhD) in discussion with the author.

308. Ibid.

309. Morris, *A Brief History of Time.*

310. Wachowski and Wachowski, *The Matrix.*

311. Ibid.

312. Ricard and Thuan, *The Quantum and the Lotus,* 178.

313. Wachowski and Wachowski, *The Matrix.*

314. Ricard and Thuan, *The Quantum and the Lotus,* 106.

315. Wachowski and Wachowski, *The Matrix.*

316. Khyentse, *The Collected Works of Dilgo Khyentse Volume Two,* 312.

317. Ibid.

318. *The Matrix Reloaded,* directed by Andy Wachowski and Larry Wachowski (Burbank, CA: Warner Brothers, 2003), DVD.

319. Ricard and Thuan, *The Quantum and the Lotus,* 33.

320. Ricard and Thuan, *The Quantum and the Lotus,* 122.

321. "A Brain in a Supercomputer," Henry Markram.

322. Khyentse, *The Collected Works of Dilgo Khyentse Volume One,* 128.

323. "A Brain in a Supercomputer," Henry Markram.

324. Ibid.

325. Ibid.

326. Ibid.

327. Ibid.

328. Khyentse, *The Collected Works of Dilgo Khyentse Volume Three,* 38.

329. "Activity-Dependent Plasticity" (n.d.), in Wikipedia, retrieved October 31, 2016 from https://en.wikipedia.org/wiki/Activity-dependent_plasticity.

330. Dale Purves et al., *Neuroscience, 4ᵗʰ ed.* (Sunderland, MA: Sinauer Associates, Inc., 2008), 625–26.

331. Ibid.

332. "Neuroplasticity" (n.d.), in Wikipedia, retrieved October 31, 2016 from https://en.wikipedia.org/wiki/Neuroplasticity.

333. Ibid.

334. Folkins et al., "Enhancing the Therapy Experience Using Principles of Video Game Design," 111–21.

335. Wachowski and Wachowski, *The Matrix.*

336. Wachowski and Wachowski, *The Matrix Revolutions.*

337. "Thalamus," S. Murray Sherman, Scholarpedia, 1, no.9:1583. Accessed January 2, 2017. doi:10.4249/scholarpedia.1583.

338. Ricard and Thuan, *The Quantum and the Lotus,* 122.

339. "Glutamate Receptor" (n.d.), in Wikipedia, retrieved September 4, 2016 from http://en.wikipeidia.org/wiki/Glutamate_receptor.

340. Ricard and Thuan, *The Quantum and the Lotus,* 33.

341. Ibid.

342. Khyentse, *The Collected Works of Dilgo Khyentse Volume Three*, 524.

343. "Black-Body Radiation," Wikipedia.

344. Wachowski and Wachowski, *The Matrix Revolutions*.

345. Khyentse, *The Collected Works of Dilgo Khyentse Volume Three*, 524.

346. "Black-Body Radiation," Wikipedia.

347. Wachowski and Wachowski, *The Matrix Revolutions*.

348. Ricard and Thuan, *The Quantum and the Lotus,* 33.

349. Khyentse, *The Collected Works of Dilgo Khyentse Volume Three*, 525.

350. *A Brief History of Time*, directed by Errol Morris.

351. Ibid.

352. Ricard and Thuan, *The Quantum and the Lotus,* 33.

353. Khyentse, *The Collected Works of Dilgo Khyentse Volume Three*, 526.

354. Ibid., 527.

355. Overbye, "No Escape from Black Holes?"

356. Ibid.

357. Khyentse, *The Collected Works of Dilgo Khyentse Volume Three*, 526.

358. Ibid.

359. Overbye, "No Escape from Black Holes?"

360. Khyentse, *The Collected Works of Dilgo Khyentse Volume Three*, 26.

361. Ibid., 33.

362. Wachowski and Wachowski, *The Matrix Reloaded*.

363. Wachowski and Wachowski, *The Matrix*.

364. Khyentse, *The Collected Works of Dilgo Khyentse Volume Three*, 29.

365. Ibid., 6.

366. *The Matrix Revolutions,* directed by Andy Wachowski and Larry Wachowski (Burbank, CA: Warner Brothers, 2003), DVD.

367. Khyentse, *The Collected Works of Dilgo Khyentse Volume Three*, 6.

368. Ibid.

369. His Holiness The Dalai Lama, *Becoming Enlightened* (New York, NY: Atria Books, 2009), 33.

370. His Holiness The Dalai Lama, *Becoming Enlightened*, 31.

371. Khyentse, *The Collected Works of Dilgo Khyentse Volume One*, 335.

372. His Holiness The Dalai Lama, *Becoming Enlightened*, 31.

373. His Holiness The Dalai Lama, *Becoming Enlightened*, 32.

374. James P. Gee, *What Video Games Have to Teach Us about Learning and Literacy* (New York: Palgrace McMillian, 2007).

375. Folkins et al., "Enhancing the Therapy Experience Using Principles of Video Game Design," 111–21.

376. Ibid.

377. Wachowski and Wachowski, *The Matrix Reloaded*.

378. Khyentse, *The Collected Works of Dilgo Khyentse Volume One*, 175.
379. Ricard and Thuan, *The Quantum and the Lotus*, 27.
380. Wallace, *The Attention Revolution*, 135.
381. Ricard and Thuan, *The Quantum and the Lotus*, 86.
382. Ricard and Thuan, *The Quantum and the Lotus*, 86.
383. Ricard and Thuan, *The Quantum and the Lotus*, 32–32.
384. Ricard and Thuan, *The Quantum and the Lotus*, 155.
385. Ricard and Thuan, *The Quantum and the Lotus*, 152.
386. Ricard and Thuan, *The Quantum and the Lotus*, 58.
387. Ricard and Thuan, *The Quantum and the Lotus*, 154.
388. Ibid
389. Overbye, "No Escape from Black Holes?"
390. Wachowski and Wachowski, *The Matrix*.
391. Wachowski and Wachowski, *The Matrix Revolutions*.
392. Ibid.
393. "Freewill and Determinism," University of Central Lancashire, Center for Professional Ethics.
394. Khyentse, *The Collected Works of Dilgo Khyentse Volume One*, 315.
395. Ibid.
396. Ricard and Thuan, *The Quantum and the Lotus*, 162.
397. Ibid.
398. Ricard and Thuan, *The Quantum and the Lotus*, 35.
399. Ricard and Thuan, *The Quantum and the Lotus*, 36.
400. Ibid
401. Ricard and Thuan, *The Quantum and the Lotus*, 20.
402. Khyentse, *The Collected Works of Dilgo Khyentse Volume Three*, 33.
403. Ibid.
404. Khyentse, *The Collected Works of Dilgo Khyentse Volume One*, 315.
405. Wachowski and Wachowski, *The Matrix Revolutions*.
406. Wachowski and Wachowski, *The Matrix Reloaded*.
407. Overbye, "No Escape from Black Holes?"
408. *The Matrix Reloaded*, directed by Andy Wachowski and Larry Wachowski (Burbank, CA: Warner Brothers, 2003), DVD.
409. Khyentse, *The Collected Works of Dilgo Khyentse Volume Two*, 265.
410. Ibid.
411. Wachowski and Wachowski, *The Matrix Reloaded*.
412. *The Matrix*, directed by Andy Wachowski and Larry Wachowski.
413. Khyentse, *The Collected Works of Dilgo Khyentse Volume Two*, 313.
414. Ibid.
415. Khyentse, *The Collected Works of Dilgo Khyentse Volume Three*, 64.
416. Ibid., 78.

417. Ibid., 69.

418. Ibid., 569.

419. Ibid., 69.

420. Khyentse, *The Collected Works of Dilgo Khyentse Volume Two*, 423.

421. Khyentse, *The Collected Works of Dilgo Khyentse Volume Three*, 224.

422. Ricard and Thuan, *The Quantum and the Lotus*, 3.

423. *It*, directed by Andy Muschietti (Burbank, CA: New Line Cinema, 2017).

424. Khyentse, *The Collected Works of Dilgo Khyentse Volume Three*, 64.

425. Wachowski and Wachowski, *The Matrix*.

426. Khyentse, *The Collected Works of Dilgo Khyentse Volume Three*, 34.

427. Ricard and Thuan, *The Quantum and the Lotus*, 20.

428. Khyentse, *The Collected Works of Dilgo Khyentse Volume Three*, 526.

ENDNOTES FOR GLOSSARY

1. "Acetylcholine" (n.d.), in Wikipedia, retrieved March 19, 2017 from https:// en.wikipedia.org/wiki/Acetylcholine.
2. "Algorithm" (n.d.), in Wikipedia, retrieved June 19, 2016 from http://wikipedia. org/wiki/Algorithm.
3. "Amygdala" (n.d.), in Merriam-Webster.com, retrieved March 19, 2017 from https://www.merriam-webster.com/dictionary/amygdala#medicalDictionary.
4. "Neuroscientist Antonio Damasio Explains Consciousness," The Huffington Post, November 16, 2010, retrieved July 24, 2016 from http://www. huffingtonpost.com.
5. "Basal Ganglia" (n.d.), in Wikipedia, retrieved March 19, 2017 from https:// en.wikipedia.org/wiki/Basal_ganglia.
6. Smythies, Edelstein, and Ramachandran, "Hypotheses Relating to the Function of the Claustrum," 53.
7. "Black hole" (n.d.), in Merriam-Webster, retrieved May 22, 2018 from https:// www.merriam webster.com/dictionary/black%20hole.
8. John Laird et al., "Claims and Challenges in Evaluating Human-Level Intelligent Systems," Proceedings of AGI-09, The Soar Cognitive Architecture, MIT Press, 2009.
9. "Brain Stem" (n.d.), in Merriam-Webster.com, retrieved March 17, 2017 from https://www.merriam-webster.com/dictionary/brain+stem.
10. "Cerebral Cortex" (n.d.), in MedlinePlus, retrieved March 19, 2017 from http://c.merriam-webster.com/medlineplus/cerebral%20cortex.
11. "Cerebral Hemisphere" (n.d.), in Merriam-Webster.com, retrieved March 19, 2017 from https://www.merriam-webster.com/dictionary/ cerebral%2Bhemisphere.
12. "Circuit" (n.d.), in Merriam-Webster.com, retrieved March 19, 2017 from https://www.merriam-webster.com/dictionary/circuit.
13. "Emergence: The Unconscious Toscanini of the Brain," Ellen Gordon, Abumrad (producers), February 18, 2005 [audio podcast] retrieved from www. radiolab.org.

14. "Cognition" (n.d.). in Merriam-Webster.com, retrieved May 22, 2018 from https://www.merriam-webster.com/dictionary/cognition.

15. "Brain" (n.d.), in Merriam-Webster.com, retrieved May 22, 2018 from https://www.merriam-webster.com/dictionary/brain.

16. "Neuroscientist Antonio Damasio Explains Consciousness," The Huffington Post, November 16, 2010, retrieved July 24, 2016 from http://www.huffingtonpost.com.

17. "Corpus Callosum" (n.d.), in Merriam-Webster.com, retrieved March 19, 2017 from https://www.merriam-webster.com/dictionary/corpus%20callosum#medicalDictionary.

18. John W. Folkins et al., "Enhancing the Therapy Experience Using Principles of Video Game Design," *American Journal of Speech-Language Pathology*, February 25, 2016, 111–21.

19. "Dopamine" (n.d.), in Wikipedia, retrieved March 19, 2017 from https://en.wikipedia.org/wiki/Dopamine.

20. "Electromagnetic radiation" (n.d.), in Merriam-Webster.com, retrieved May 22, 2018 from https://www.merriam-webster.com/dictionary/electromagnetic%20radiation

21. Damasio, *The Feeling of What Happens. Body and Emotion in the Making of Consciousness*, 50.

22. Ibid.

23. "Electromagnetic Radiation," Wikipedia.

24. "Elementary Particle" (n.d.), in Wikipedia, retrieved March 19, 2017 from https://en.wikipedia.org/wiki/Elementary_particle.

25. Folkins et al., "Enhancing the Therapy Experience Using Principles of Video Game Design," 111–21.

26. "Forebrain" (n.d.), in MedlinePlus, retrieved March 20, 2017 from http://c.merriam-webster.com/medlineplus/forebrain.

27. John Holland, *Adaptation in Natural and Artificial Systems*, (Ann Arbor, Michigan, U. Michigan Press, 1975.) J. J. Hopfield, "Neural Networks and Physical Systems with Emergent Collective Computational Properties," *Proc. Nat. Acad. Sci.* (USA) 79, (1982) 2554–2558.

28. "Glutamate Receptor" (n.d.), in Wikipedia, retrieved September 4, 2016 from http://en.wikipeidia.org/wiki/Glutamate_receptor.

29. "Visualization of Glutamate as a Volume Transmitter," *J Physiol.*, 2011 Feb 1;589 (Pt. 3):481–8. doi: 10.1113/jphysiol.2010.199539. Epub 2010 Nov 29.

30. Ibid.

31. "Gray Matter" (n.d.), in Merriam-Webster.com, retrieved March 19, 2017 from https://www.merriam-webster.com/dictionary/gray%20matter.

32. Barbara Hammer and Pascal Hitzler, eds., "Perspectives of Neural-Symbolic Integration," *Studies in Computational Intelligence*, Vol. 77. Springer, 2007.

33. "Hypothalamus" (n.d.), in MedlinePlus, retrieved March 19, 2017 from http://c.merriam-webster.com/medlineplus/hypothalamus

34. L. W. Swanson, *Brain Architecture: Understanding the Basic Plan* (New York: Oxford University Press, 2003).

35. Goertzel, "Artificial Intelligence."

36. James P. Gee, *What Video Games Have to Teach Us about Learning and Literacy* (New York: Palgrace McMillian, 2007).

37. "Measuring and Defining the Experience of Immersion in Games," Charlene Jennett et al.

38. Goertzel, "Artificial Intelligence."

39. Hammer and Hitzler, "Perspectives of Neural-Symbolic Integration."

40. "Intelligence" (n.d.), in Merriam-Webster.com, retrieved May 22, 2018 from https://www.merriam-webster.com/dictionary/intelligence

41. "Plasticity" (n.d.), in Merriam-Webster.com, retrieved May 22, 2018 from https://www.merriam-webster.com/dictionary/plasticity.

42. Smythies, Edelstein, and Ramachandran, "Hypotheses Relating to the Function of the Claustrum,"

43. Milardi et al., "Cortical and Subcortical Connections of the Human Claustrum Revealed in Vivo by Constrained Spherical Deconvolution Tractography," 406–14.

44. "Causality" (n.d.), in Merriam-Webster.com, retrieved May 22, 2018 from https://www.merriam-webster.com/dictionary/causality.

45. Khyentse, *The Collected Works of Dilgo Khyentse Volume Two*, 112.

46. Milardi et al., "Cortical and Subcortical Connections of the Human Claustrum Revealed in Vivo by Constrained Spherical Deconvolution Tractography," 406–14.

47. Khyentse, *The Collected Works of Dilgo Khyentse Volume Two*, 112.

48. "Motor Neuron" (n.d.), in Wikipedia, retrieved March 20, 2017 from https://en.wikipedia.org/wiki/Motor_neuron.

49. "Neocortex" (n.d.), in Merriam-Webster.com, retrieved May 22, 2018 from https://www.merriam-webster.com/dictionary/neocortex

50. Markram, "A Brain in a Supercomputer."

51. Goertzel, "Artificial Intelligence."

52. Markram, "A Brain in a Supercomputer."

53. "Norepinephrine" (n.d.), in Wikipedia, retrieved March 19, 2017 from https://en.wikipedia.org/wiki/Norepinephrine.

54. "Dark Energy, Dark Matter," NASA Science, accessed July 24, 2016, http://science.nasa.gov/astrophysics/focus-areas/what-is-dark-energy.

55. "Nucleus (Neuroanatomy)" (n.d.), in Wikipedia, retrieved March 19, 2017 from https://en.wikipedia.org/wiki/Nucleus_(neuroanatomy).

56. "Photon" (n.d.), in Wikipedia, retrieved March 19, 2017 from https://en.wikipedia.org/wiki/Photon.

57. Milardi et al., "Cortical and Subcortical Connections of the Human Claustrum Revealed in Vivo by Constrained Spherical Deconvolution Tractography," 406–14.

58. Khyentse, *The Collected Works of Dilgo Khyentse Volume Two*, 112.

59. "Neuroscientist Antonio Damasio Explains Consciousness," The Huffington Post.

60. Ibid.

61. "Quantum," (n.d.), in Wikipedia, retrieved March 19, 2017 from https://en.wikipedia.org/wiki/Quantum.

62. "Reality" (n.d.), in Merriam-Webster.com, retrieved May 23, 2018 from https://www.merriam-webster.com/dictionary/reality.

63. "Serotonin" (n.d.), in Wikipedia, retrieved March 19, 2017 from https://en.wikipedia.org/wiki/Serotonin.

64. "Cognition and Emotion," Pessoa.

65. "Spinal Cord" (n.d.), in Wikipedia, retrieved March 20, 2017 from https://en.wikipedia.org/wiki/Spinal_cord.

66. Milardi et al., "Cortical and Subcortical Connections of the Human Claustrum Revealed in Vivo by Constrained Spherical Deconvolution Tractography," 406–14.

67. Khyentse, *The Collected Works of Dilgo Khyentse Volume Two*, 112.

68. "Thalamus," S. Murray Sherman, Scholarpedia, 1, no.9:1583. Accessed January 2, 2017. doi:10.4249/scholarpedia.1583.158. Vertebrate(n.d.) In Merriam-Webster.com. Retrieved March 20, 2017 from https://www.merriam-webster.com/dictionary/vertebrate#h2

69. "Vertebrate" (n.d.), in Merriam-Webster.com, retrieved May 23, 2018 from https://www.merriam-webster.com/dictionary/vertebrate.

ENDNOTES FOR APPENDIX A

1. Antonio Damasio, The Feeling of What Happens. Body and Emotion in the Making of Consciousness (New York: Harcourt, Inc., 1999), 26.

2. Andrea D. Grabovac, Mark A. Lau and Brandilyn Willett, "Mechanisms of Mindfulness: A Buddhist Psychological Model," Springer Science+Business Media, LLC. (2011).

3. Grabovac, Lau and Willett, "Mechanisms of Mindfulness: A Buddhist Psychological Model."

4. "More or Less Human," Simon Adler (producer), May 17, 2018 [audio podcast] retrieved from https://www.wnycstudios.org/story/more-or-less-human/.

5. "More or Less Human," Simon Adler (producer).

6. "The Row," Pat Mesiti-Miller, Earlonne Woods and Nigel Poor (producers), May 23, 2018, [audio podcast] retrieved from https://www.earhustlesq.com/episodes/2018/5/23/the-row.

7. Dilgo Khyentse, The Collected Works of Dilgo Khyentse Volume One (Boston: Shambhala Publications, Inc., 2010), 317.

8. Khyentse, The Collected Works of Dilgo Khyentse Volume Two, 455.

9. Khyentse, The Collected Works of Dilgo Khyentse Volume Two, 260.

10. "A Brain in a Supercomputer," Henry Markram, TED video, filmed July 2009, posted April 15, 2009, http://www.ted.com/.

11. Khyentse, The Collected Works of Dilgo Khyentse Volume Two, 260.

ENDNOTES FOR APPENDIX B

1. *The Matrix*, directed by Andy Wachowski and Larry Wachowski (Burbank, CA: Warner Brothers, 1999), DVD.
2. Wachowski and Wachowski, *The Matrix*.
3. Wachowski and Wachowski, *The Matrix*.
4. Wachowski and Wachowski, *The Matrix*.
5. Wachowski and Wachowski, *The Matrix*.
6. Wachowski and Wachowski, *The Matrix*.
7. Wachowski and Wachowski, *The Matrix*.
8. Wachowski and Wachowski, *The Matrix*.
9. Antonio Damasio, *The Feeling of What Happens. Body and Emotion in the Making of Consciousness* (New York: Harcourt, Inc., 1999), 198.
10. Wachowski and Wachowski, *The Matrix*.
11. Wachowski and Wachowski, *The Matrix*.
12. Wachowski and Wachowski, *The Matrix*.
13. Wachowski and Wachowski, *The Matrix*.
14. Wachowski and Wachowski, *The Matrix*.
15. Wachowski and Wachowski, *The Matrix*.
16. Wachowski and Wachowski, *The Matrix*.
17. Wachowski and Wachowski, *The Matrix*.
18. Wachowski and Wachowski, *The Matrix*.
19. Wachowski and Wachowski, *The Matrix*.
20. Wachowski and Wachowski, *The Matrix*.
21. Wachowski and Wachowski, *The Matrix*.
22. Wachowski and Wachowski, *The Matrix*.
23. Wachowski and Wachowski, *The Matrix*.
24. *The Matrix Reloaded*, directed by Andy Wachowski and Larry Wachowski (Burbank, CA: Warner Brothers, 2003), DVD.
25. Wachowski and Wachowski, *The Matrix Reloaded*.
26. Matthieu Ricard and Trinh Xuan Thuan, *The Quantum and the Lotus* (New York, New York: Three Rivers Press, 2001), 33.
27. Wachowski and Wachowski, *The Matrix Reloaded*.

28. Ricard and Thuan, *The Quantum and the Lotus,* 27.

29. B. Alan Wallace, *The Attention Revolution* (Somerville, MA: Wisdom Publications, Inc., 2006), 138.

30. Wachowski and Wachowski, *The Matrix Reloaded.*

31. "How Many Neurons Make a Human Brain? Billions Fewer than We Thought," James Randerson, *The Guardian*, February 28, 2012, accessed March 16, 2016, http://www.theguardian.com.

32. John Smythies, Lawrence Edelstein, and Vilayanur Ramachandran, "Hypotheses Relating to the Function of the Claustrum," *Frontiers in Integrative Neuroscience* 6 (August 2012): 53.

33. Wachowski and Wachowski, *The Matrix Reloaded.*

34. Khyentse, *The Collected Works of Dilgo Khyentse Volume Three*, 29.

35. Khyentse, *The Collected Works of Dilgo Khyentse Volume Three*, 6.

36. Ibid.

37. Wachowski and Wachowski, *The Matrix Reloaded.*

38. "Neuroscientist Antonio Damasio Explains Consciousness," The Huffington Post.

39. Ibid.

40. Damasio, *The Feeling of What Happens. Body and Emotion in the Making of Consciousness*, 50.

41. "Neuroscientist Antonio Damasio Explains Consciousness," The Huffington Post.

42. Wachowski and Wachowski, *The Matrix Reloaded.*

43. "Freewill and Determinism," University of Central Lancashire, Center for Professional Ethics, accessed June 20, 2016, http://uclan.ac.uk.

44. *Fentanyl: The Drug Deadlier than Heroin*, VICE, published on July 22, 2016, accessed September 30, 2016.

45. Wachowski and Wachowski, *The Matrix Reloaded.*

46. Barbara Hammer and Pascal Hitzler, eds., "Perspectives of Neural-Symbolic Integration," *Studies in Computational Intelligence*, Vol. 77. Springer, 2007.

47. "Neural Correlates of Consciousness," Mormann Florian and Christof Koch, Scholarpedia 2, no.12(2007):1740. Accessed August 20, 2016. doi:1.4249/scholarpedia.1740.

48. "Emotional Memory," Joseph E. LeDoux, Scholarpedia 2, no. 7: (2007): 2698. Accessed June 22, 2016. doi:10.4249/scholarpedia.1806.

49. Wallace, *The Attention Revolution,* 121.

50. Wallace, *The Attention Revolution,* 138.

51. Wachowski and Wachowski, *The Matrix Reloaded.*

52. Wachowski and Wachowski, *The Matrix Reloaded.*

53. Wachowski and Wachowski, *The Matrix Reloaded.*

54. Dennis Overbye, "No Escape from Black Holes? Stephen Hawking Points to a Possible Exit," *New York Times*. Last modified June 2016.https://www.nytimes.com/2016/06/07/science/stephen-hawking-black-holes.html?_r=0.

55. *The Matrix Revolutions,* directed by Andy Wachowski and Larry Wachowski (Burbank, CA: Warner Brothers, 2003), DVD.

56. Wachowski and Wachowski, *The Matrix Revolutions.*

57. Wachowski and Wachowski, *The Matrix Revolutions.*

58. Dilgo Khyentse, *The Collected Works of Dilgo Khyentse Volume Three* (Boston: Shambhala Publications, Inc., 2010), 26.

59. Wachowski and Wachowski, *The Matrix Revolutions.*

60. Wachowski and Wachowski, *The Matrix Revolutions.*

61. Wachowski and Wachowski, *The Matrix Revolutions.*

62. Wachowski and Wachowski, *The Matrix Revolutions.*

63. Wachowski and Wachowski, *The Matrix Revolutions.*

64. John W. Folkins et al., "Enhancing the Therapy Experience Using Principles of Video Game Design," *American Journal of Speech-Language Pathology,* February 25, 2016, 111–21.

65. Wachowski and Wachowski, *The Matrix Revolutions.*

ABOUT THE AUTHOR

Holly Marie Pollard-Wright has spent her life seeking to understand the nature of reality. With a firm grasp of neurology, cognitive science, and theoretical physics as a foundation, she began to explore the teachings of Tibetan master and scholar Dilgo Khyentse Rinpoche. She is a practicing veterinarian, and her husband, Mark, is a medical doctor. They have a superhero son, Benny.

www.ingramcontent.com/pod-product-compliance
Lightning Source LLC
Chambersburg PA
CBHW031823170526
45157CB00001B/161